NONDESTRUCTIVE TESTING OF DEEP FOUNDATIONS

NONDESTRUCTIVE TESTING OF DEEP FOUNDATIONS

Bernard Hertlein
STS Consultants Ltd, USA

and

Allen Davis
Construction Technology Labs Inc., USA

John Wiley & Sons, Ltd

Telephone (+44) 1243 779777

Email (for orders and customer service enquiries): cs-books@wiley.co.uk
Visit our Home Page on www.wiley.com

Other Wiley Editorial Offices

John Wiley & Sons Inc., 111 River Street, Hoboken, NJ 07030, USA

Jossey-Bass, 989 Market Street, San Francisco, CA 94103-1741, USA

Wiley-VCH Verlag GmbH, Boschstr. 12, D-69469 Weinheim, Germany

John Wiley & Sons Australia Ltd, 42 McDougall Street, Milton, Queensland 4064, Australia

John Wiley & Sons (Asia) Pte Ltd, 2 Clementi Loop #02-01, Jin Xing Distripark, Singapore 129809

John Wiley & Sons Canada Ltd, 22 Worcester Road, Etobicoke, Ontario, Canada M9W 1L1

Wiley also publishes its books in a variety of electronic formats. Some content that appears in print may not be available in electronic books.

Library of Congress Cataloging-in-Publication Data

Hertlein, Bernard H.
Nondestructive testing of deep foundations / Bernard H. Hertlein, and Allen G. Davis.
p. cm.
Includes bibliographical references and index.
ISBN 0-470-84850-2 (cloth : alk. paper)
1. Foundations–Testing. 2. Piling (Civil engineering)–Testing. 3. Nondestructive testing.
I. Davis, Allen G. II. Title.

TA775.H395 2006
624.1′50287–dc22

2006010653

A catalogue record for this book is available from the British Library

ISBN-13 978-0-470-84850-0 (HB)
ISBN-10 0-470-84850-2 (HB)

Typeset in 10/12pt Times by TechBooks, New Delhi, India.

This book is printed on acid-free paper responsibly manufactured from sustainable forestry in which at least two trees are planted for each one used for paper production.

Contents

Foreword

Nondestructive testing (NDT) of deep foundations has become an essential component of deep foundation construction *quality assurance*. Its very existence has improved the front end of the process, that being *quality control*.

This is a text whose time has more than come. The authors' comprehensive treatment of a complex and sometimes obscure subject has the potential to render the careful reader quite well informed and the almost lyrical style in which it is written only serves to enhance that possibility. A word of caution however – while a well-trained and experienced technician may be capable of performing the field work for most NDT techniques, supervision and final interpretation of NDT data generally require the expertise of an experienced engineering professional. Hertlein and Davis are infinitely qualified to have taken on this project; both are highly respected NDT theorists and practitioners. With regard to the interest and concerns of the deep foundation construction industry their 'practitioner' experience is most important. The authors take a no-nonsense approach to a subject that often lends itself to theoretical esoteria, which is not the construction industry's strong suit. Their objective and critical treatment of the advantages and disadvantages of a wide variety of testing techniques is evidence of their unique sensitivity to the sometimes-arcane science of NDT. This work is an important contribution to the literature and more importantly to the practicing civil engineering and civil construction communities.

The deep foundation industry is in the authors' debt for their taking on this daunting project.

S. Scot Litke, Executive Director
ADSC: The International Association of Foundation Drilling

Preface

Both authors of this book have been involved in the development and application of nondestructive test techniques for deep foundations for a very large part of their lives and have worked in many parts of the world. In doing so, they have concurrently observed and contributed to improvements in deep foundation design, construction, inspection and testing techniques, and developments in construction equipment and materials. As in all industries where advances occur almost daily, and are driven by a combination of technological interest and financial need, occasional mistakes are made, and today's miracle solution sometimes becomes tomorrow's lesson for the hasty.

Like all other forms of nondestructive or non-invasive testing, nondestructive testing of deep foundations is the subject of many myths and legends. In reality, every test method has specific capabilities and limitations that govern the usefulness of the test results. While every test technique in commercial use has proven capabilities, the results of any test used in the wrong circumstances will, at best, be inconclusive, and at worst may be downright misleading.

The initial aim of this book was to provide a starting point and logical guide for geotechnical and/or structural engineering students with an interest in deep foundations and a useful reference for practicing professionals who have never been exposed to testing techniques for deep foundations, but, through no fault of their own, suddenly find themselves needing to specify or hire such testing services. However, as the book evolved, so did the authors' appreciation of the common misconceptions in the deep foundations industry.

The need for this book was illustrated recently by a project in which the engineer was relying on a report published by the FHwA in 1993 (Baker *et al.*). A problem had occurred during the construction of a drilled shaft, which gave rise to the possibility of soil contamination of the concrete about halfway down the shaft. A surface reflection technique such as the Impulse-Response test was suggested, but the project engineer refused to accept that proposal on the grounds that the FHwA report stated that such techniques could only detect major defects in the shaft. While that may have been the case at the time of writing the FHwA report, both the techniques and the skills of the people applying them have improved steadily over the last decade or so. In the present authors' experience, the effectiveness of surface reflection techniques is governed by several factors, including shaft length/diameter ratio and the depth to the anomaly. Accurate location of anomalies that affect 15 % or less of the shaft cross-section is not uncommon. By sticking to a rigid application of the opinions given in the FHwA report, the engineer was refusing to acknowledge that the techniques and the practitioners' skills have improved since that report was written. That is not to say that the FHwA report is somehow flawed – in the present authors' opinions, it is an accurate reflection of the state-of-the-art at the time it was written. Hopefully, however, this book will serve to remind potential users of NDT methods that in the decade since the FHwA report was published, the technology and the expertise associated with it have steadily improved, and will continue to do so.

By including a history of the various deep foundation testing techniques, and a comprehensive list of references, the authors hope that they have provided all who read this book the opportunity to gain an understanding of the reasons for the various techniques, an appreciation of the capabilities of the various methods, an understanding of why there can be no 'one-size-fits-all' technique for testing deep foundations and a resource for finding additional information. At the same time, we have endeavored to make this book as up-to-date as possible, while recognizing the dynamic nature of nondestructive testing. This sector of the deep foundations industry is blessed with many innovative and enquiring minds and so, hopefully, will be in a constant state of exploration and development for many years to come – thus, dear reader, in the following pages you will come across the phrase 'at the time of writing', or something similar, quite often. This book is our best attempt at capturing the state of the art 'at the time of writing' – but expect it to change soon!

Bernard H. Hertlein and Allen G. Davis

About the Authors

BERNARD H. HERTLEIN, M.ASCE

Bernard Hertlein started his professional career as a mechanical engineering student but soon realized that he had a natural affinity for electronics. After working professionally with automotive electronic systems and pursuing audio engineering as a hobby for several years, he migrated through audio engineering to instrumentation of civil engineering structures and finally to nondestructive testing. After joining Testconsult, the English subsidiary of the French National Center for Building and Civil Engineering Research (CEBTP), Mr Hertlein became deeply involved with the development of both the software and the hardware for several nondestructive test techniques for deep foundations that are now in common use worldwide. He worked on construction testing projects throughout Europe, Hong Kong, parts of North Africa and the United States, eventually settling in the United States, where he and Allen Davis introduced the Cross-hole Sonic-Logging technique, the Parallel-Seismic test and the Impulse-Response (Sonic-Mobility) test in the mid 1980s.

In 1992, Mr Hertlein joined STS Consultants, based in Vernon Hills, Illinois, where he has continued to design and build test equipment and research new applications for the test techniques that he had helped to introduce to the United States. Mr Hertlein is an active member of several key professional societies. At the time of writing this book, he is a member of the American Society of Civil Engineers, Chairman of the Nondestructive and In-place Testing Committee of the American Society for Testing and Materials (ASTM C9-64), Chairman of the Testing and Evaluation Committee of the Deep Foundations Institute, and Secretary of the Nondestructive Testing Committee of the American Concrete Institute (ACI 228). He also serves as a member of ACI Committee 336: Footings, Mats and Drilled Piers, ASTM Committee C9-47: Self-Consolidating Concrete, D18-11: Deep Foundations and G9-14: Corrosion of Reinforcing Steel.

Mr Hertlein has written numerous conference papers and journal articles. He is a regular member of the faculty for the International Association of Foundation Drilling (ADSC-IAFD) Drilled Shaft Inspector's School and a frequent lecturer at other educational seminars presented by ACI International, ADSC regional chapters, the ASCE Geo-Institute and the Deep Foundations Institute.

ALLEN G. DAVIS, PH.D., D.SC., PE

Allen Davis qualified as a geologist and his first career was as prospector for De Beers Corporation in Central Africa. He then converted to Civil Engineering through Geotechnics, gaining his Ph.D. from Birmingham University, UK in that subject. He has had Academic, Research and Industrial experience in fairly equal proportions, including:

- Professor at the University of Birmingham, UK, for 10 years.
- Head of the Geotechnical and Highways Research Division, National Center for Building and Civil Engineering Research (CEBTP), Paris, France, for 8 years.
- Technical and Managing Director, Testconsult CEBTP (UK) for 8 years. He was one of the founding members of Testconsult in 1974.
- Principal Engineer, STS Consultants, Ltd and Manager for NDE, Madsen, Kneppers & Associates, Chicago, Illinois and Salt Lake City, Utah, USA for 6 years.
- Senior Principal Engineer, Construction Technology Laboratories, Inc. (CTL), Skokie, Illinois, USA for the last 6 years.

At the time of writing this book he was Manager of Nondestructive Evaluation at CTL in Skokie, Illinois. His special interests included vibration problems and real-time data acquisition from dynamic testing of concrete foundations and structures, and he was a member and past Chairman of Committee 228 (Nondestructive Testing of Concrete) of the American Concrete Institute and also a member of ASTM Committee C9-64 – Nondestructive and In-place Testing. He has published over 80 technical articles and publications to date in the fields of Civil Engineering and Building, Transportation and Materials Resources. Eleven Ph.D. research students (seven in France, four in England) have graduated under his supervision, and he was awarded the degree of Doctor of Science by Birmingham University in 1980.

His contributions to the concrete industry and to the work of the ACI were recognized at the October 2004 meeting of the ACI in San Francisco, where it was announced that he had been elected a Fellow of the Institute. Unfortunately ill-health had prevented him from going to San Francisco, and he passed away suddenly at his home a few hours after learning of the fellowship announcement. Rest in peace, old chum.

Acknowledgements

Trying to write a state-of-the art summary for an industry that is global in reach but at the same time is restricted by national and sometimes even regional codes and practices is a task that is beyond the capabilities of one or two people. The authors therefore gratefully acknowledge the assistance of industry experts around the world who have contributed to this book by giving permission to use quotations and/or illustrations, offering constructive criticism, or just simply let us pick their brains in conversations and professional committee activities, and with that wonderful tool of the modern researcher – the Internet.

Thanks are due to our families and work colleagues for their forbearance at those times when we were preoccupied with this work – discussing chapters over lunch or dinner or hunting through our respective office libraries to verify reference material. For this latter effort, we sometimes even resorted to ransacking a colleague's office bookshelf!

The list of people to whom special thanks are due illustrates perfectly the global reach and international importance of nondestructive testing for deep foundations. Specific contributions or suggestions that have been incorporated in this book have come from direct communication and/or participation in professional committees activities by the following:

Joram Amir of PileTest, Israel
Zaw Zaw Aye of SEAFCO, Bangkok, Thailand
Clyde Baker Jr of STS Consultants, Illinois, USA
Jorge Beim of PDI Engenharia, Rio de Janeiro, Brazil
Gordon Cameron of Napier University, Edinburgh, UK
Geordie Compton of the Deep Foundations Institute, New Jersey, USA.
Conrad Felice of Lachel Felice Company, Washington, USA
Bengt Fellenius of Alberta, Canada
Hernan Goldemberg of Geotecnica Cientec SA, Buenos Aires, Argentina
Carl-John Grävare of Pile Dynamics Europe AB, Sweden

Mike Holloway of Insitutech, California, USA
Alain Holeyman of the University of Louvain, Belgium
Mike Justason of Berminghammer Corporation, Ontario, Canada
Henk van Koten, formerly of TNO, Holland
Ted Ledgard of the International Association of Foundation Drilling, Texas, USA
Garland Likins of Pile Dynamics, Inc., Ohio, USA
K. Rainer Massarsch, Consultant, Sweden
Peter Middendorp of Profound, Holland
Mike Muchard of Applied Foundation Testing, Inc., Florida, USA
Claus Peterson of Germann Instruments, Illinois, USA and Denmark
Mark Presten of American Piledriving, Inc., California, USA
Frank Rausche of GRL Engineers, Ohio, USA
David Redhead of BSP International Foundations, UK
Julian Seidel of Monash University, Victoria, Australia
Dick Stain of Testconsult Ltd, Risley, UK
Greg Wennerstrom of ENMAX, Alberta, Canada
Huw Williams of Testconsult, Ltd, Risley, UK

This is an incomplete list, selected only to show the breadth of national and international participation. In all fairness, most members of the professional committees of the International Association for Foundation Drilling and DFI International, plus the deep foundation committees of the ASCE Geo-Institute and ASTM, have had some influence on the contents of this book, but to list them all would read as an elite *Who*'s *Who* of the deep foundations industry and take up too much space here. The fact that they are active members of those organizations by serving on the various committees testifies to their commitment to the deep foundations industry and to the quality of their product.

Thank you, ladies (yes, there are some real ladies in this business!) and gentlemen, for helping us to document the international importance of NDT for deep foundations. The penultimate thank you goes to Scott Litke, for finding the time to write the Foreword. As executive director of the International Association of Foundation Drilling, Scott has been instrumental both in keeping NDT practitioners on their toes and in helping foundation contractors realize and take advantage of the benefits offered by NDT.

Finally – thank you to Wendy Hunter, Kelly Board and Sarah Powell of John Wiley & Sons, Ltd, Chichester, UK for your patience and guidance in a project that took quite a while longer than we anticipated, but we feel was well worth the effort.

Photography and Illustration Credits

Thanks are also due to the individuals and companies who gave us permission to use some of their photographs or graphics to illustrate this book. The following photographs or illustrations have been reproduced by kind permission of the following owners:

- The illustrations of the pile and casing vibrator in Chapter 2, Figures 2.3, 2.4(a) and 2.4(b) – John White of American Piledriving Equipment, Washington, USA.
- The photographs of press-in pile installation systems in Chapter 2, Figures 2.5 and 2.10 – Michael Wysockey of Thatcher Engineering, Illinois, USA.
- The illustration of the stages of auger-cast pile construction in Chapter 2, Figure 2.8 – West Coast Foundation, Florida, USA.
- The photographs of the rogue driven pile in Chapter 4, Figures 4.1 and 4.2 – Dr Bengt Fellenius, of Urkkada, Inc., Alberta, Canada.
- The photograph of the shaft inspection device in Chapter 4, Figure 4.5 and the photographs of the Statnamic load-test equipment in Chapter 6, Figures 6.12–6.17 – Mike Muchard of Applied Foundation Testing, Inc., Florida, USA.
- The wire-frame model images from the ultrasonic borehole caliper in Chapter 4, Figure 4.6 – Matt Sutton of R&R Visuals, Inc., Indiana, USA.
- The graphic illustration of wave propagation in a deep foundation in Chapter 6, Figures 6.1(a) and 6.1(b) – Peter Middendorp of Profound, BV, The Netherlands.
- The photograph of 'Newton's Apple' in Chapter 6, Figure 6.11 – Garland Likins of GRL Engineers, Inc., Ohio, USA.
- The photographs of the Fundex pile-load tester in Chapter 6, Figures 6.18–6.20 – Mark Presten of American Piledriving, Inc., California, USA.

- The three-dimensional tomography images in Chapter 10, Figures 10.7 and 10.8 – Huw Williams of Testconsult, Ltd, Risley, UK.
- The photographs of concrete taken through a fiber optic device in Chapter 13, Figures 13.8–13.10 – Mahmood Samman of Stress Engineering, Texas, USA.

1

Introduction and a Brief History

1.1 INTRODUCTION

Nondestructive Testing (NDT) for deep foundations is very much an expression of the state-of-the-art of the electronics and computer industries, materials science and our ability to bring them together and put them all to practical use. All three of these crucial areas have seen tremendous advances in the last ten years. This book is not intended to supplant any of its forebears, but rather to build on their foundation (no pun intended!) by reminding the reader of the origins of the techniques (and the assumptions made at the time!) and bringing the reader up-to-date with the enormous gains that our industry has recently made; hence, a suitable subtitle for this book – 'Another Decade of Technical Advances'. The decade in question started with the publication of the Federal Highway Administration (FHwA) report 'Drilled Shafts for Bridge Foundations' (Baker *et al.*, 1993). This report was the essential key to acceptance of NDT techniques for drilled shafts and augered, cast-in-place (ACIP) piles in the United States, which in turn increased the acceptance and use of both drilled shafts and ACIP piles as reliable foundation techniques. It is a fact that engineers in many countries were schooled in the USA and look to the engineering community in the USA for guidance. The FHwA report thus had a significant effect on both the testing community and the deep foundations industry worldwide. It is, at the time of writing this book, still the international benchmark for many testing specifications.

Nature has regularly taken its toll of the works of man through unexpected catastrophes, ranging from flooding to earthquakes to volcanic eruptions. Rivers in flood scour away the soil supporting bridge piers – sands beneath high-rise buildings behave like a liquid when an earthquake strikes – glaciers on high volcanoes melt in minutes during an eruption, triggering devastating mudslides that spread for miles. Each natural event reminds us that the stability of the soil cannot be taken for granted

Nondestructive Testing of Deep Foundations B.H. Hertlein and A.G. Davis

and that deep foundations must often be designed to do much more than to simply support the mass of the structure built upon them.

In addition to this, population growth and commercial expansion create pressures that demand higher buildings and larger structures with each succeeding generation. This increases the need for deep foundations, often in less than ideal geotechnical and physical conditions. It is to the credit of civil engineers and the deep foundation construction industry that they have always found ways to meet these demands, often using innovative designs, construction techniques and materials. So much innovation is not without its risks, however, and foundation failures have occurred because quality control techniques failed to keep pace with deep foundation technology. Having said that, it must also be acknowledged that it is often extremely difficult to distinguish between foundation failure due to defects in the foundation itself, and failure of the surrounding soil or bearing strata. Notable examples are discussed in more detail later in this chapter, under the subheading 'Deep Foundation Failures and NDT'. Unfortunately, corruption and deliberate malfeasance are also sometimes factors in the creation of substandard foundations – the nature of the problem is then indisputable, but the actual cause becomes clouded in 'finger-pointing' and legalities.

Thus, the forces of nature and the needs of mankind have created a demand for both deep, stable foundations, and for quality control techniques that can ensure their reliability. Proof testing of each and every foundation by static loading to twice the maximum probable seismic or catastrophic load is a practical and financial impossibility, yet owners and engineers alike are reluctant to accept something that they cannot see without some form of assurance that it is sound, and will perform as designed. It is a basic truth that 'necessity is the mother of invention' – and never more so than in this case. Non-invasive and/or nondestructive alternatives to full-scale tests were developed specifically to assess a foundation's integrity and/or predict likely performance without raising project costs to prohibitive levels.

NDT of deep foundations is a complex topic covering a number of different techniques designed to gain information about the integrity and quality of the material that makes up a deep foundation. Typical foundation materials are concrete, timber, steel and rock. Deep foundations vary in size and shape, may be constructed of a combination of materials and may be built by a combination of several different techniques. Each combination of size, shape, material and construction method creates a unique set of circumstances that includes the risk of a variety of defects specific to those circumstances. Those same circumstances will determine the accessibility of the foundation for inspection during construction and for NDT examination after construction.

The variations in possible defect types, foundation access and construction material have led to the development of several different NDT methods over the last 30 years. Each method has been designed for a specific purpose and a defined range of circumstances, and therefore has a specific and unique set of capabilities that determine its applicability to a particular project. Conversely, each method also has a specific set of limitations that may adversely affect its effectiveness or the reliability of the data generated under certain circumstances. Using a test method that is inappropriate for a given set of circumstances or for the information that is being sought will, at best, be inconclusive, and at worst may be actively misleading (Stain, 1982).

In order to be able to specify an appropriate method or to recognize an inappropriate specification, it is necessary for the engineer, specifier and/or contractor to not only understand the capabilities and limitations of each of the methods in use today, but also to be aware of the potential problems for both construction and testing that are inherent in each type of foundation and in the local soils. This manual therefore describes the most commonly used deep foundation construction techniques, the limitations imposed by the local soils, typical use of materials and the NDT methods commercially available at the time of writing. It also aims to increase the reader's understanding of these factors by providing an overview of the principle types of deep foundation, a brief history of the development of NDT, a description of the various NDT methods and a summary of the capabilities and limitations of each method.

1.2 A BRIEF HISTORY OF DEEP FOUNDATIONS AND THE ADVENT OF NDT

1.2.1 CAVEAT AND ACKNOWLEDGEMENT

The authors of this book recently participated in the writing of a manual for the inspection of drilled shafts, sponsored by the Deep Foundations Institute. This work included the preparation of a brief history of both high- and low-strain tests. During the course of researching and summarizing the history of these methods, it became apparent that most histories of engineering are, sadly, colored by the culture of the historian in much the same way that the histories of conflicts are colored by the side with which the reporter sympathized.

We are, today, spoiled by the ease of access to knowledge that the Internet provides – it sometimes makes us forget that not all knowledge is yet available 'at the click of a mouse'. Researchers before the 1980s had no Internet or Worldwide Web and had to rely on old-fashioned footwork and laborious library searches. Sterling work was often published in obscure local or national society publications and rarely received international recognition; being digitized and posted on 'the Web' was never even an option. Such work was also often written in a language unfamiliar to other researchers. Small wonder, then, that researchers in any country might be unaware of the achievements of colleagues in other countries and were therefore doomed to 'reinvent the wheel' on numerous occasions.

The history reported in this book is a result of the best efforts of the authors and numerous reputable sources to keep the facts straight and unbiased by commercial interests. Much of the research was performed in the old-fashioned way – by personal recall, interview of industry veterans and by library searching. We apologize for any omissions or oversights, which we must attribute to gaps in the collective knowledge of the industry. By the same token, we do not apologize for any similarities between the history published here and the history published in the DFI manual – history should be history, no matter who recounts it!

1.2.2 THE HISTORY

Since the dawn of civilization, Mankind has been aware of the need for a stable foundation if any substantial structure is to survive for long without settling, cracking or sometimes just falling apart! Often, simply digging through soft topsoil to a stiffer underlying soil and piling up rocks proved adequate. In other cases, particularly in deep sand or alluvial soils, deeper foundations were needed. Nobody really knows who first came up with the idea of stripping a tree-trunk and banging it into the ground, but driven timber piles have been around almost as long as people have lived in constructed homes. When the ancient Roman Empire was at the height of its splendor, driving of timber piles was already regarded a documented science. Recent historical research sponsored by American Piling Equipment was published by the Deep Foundations Institute in the *Deep Foundations* magazine (Smith, 2005). According to Smith, the oldest bridge built by the Romans was the *Pons Publicius* (Bridge of Piles), constructed by Ancus Martius in or about 621 BC. Unfortunately, Smith is not clear as to whether it was truly the oldest Roman bridge or merely the oldest Roman bridge still in existence.

While some of the 'science' appears primitive and flawed by current standards, Smith shows that the subject of timber pile foundations had been thoroughly examined and documented by Marcus Vitruvius Pollio (Vitruvius) in his *De Architectura, The Ten Books on Architecture*, believed to have been written between 27 and 23 BC. Vitruvius not only wrote about the appropriate design of timber piled foundations, but described the proper method of harvesting the timber to ensure longevity, and discussed why some species of timber last longer than others. His theories as to the cause of the difference in rot resistance of the various species may appear laughable to a modern scientist, but demonstrate that the Romans were clearly aware of the fact that timber driven below the ground water table lasts considerably longer than timber exposed to the atmosphere. Julius Caesar, in his book on the Gallic wars, *De Bello Gallico,* described the installation of inclined piles to resist river current forces during the construction of a bridge over the River Rhine near Koblenz in 55 BC. The reader interested in the history of driven piles before the electronic era will find many gems of knowledge about the subject handily condensed in Smith's article.

In more recent times, greater loads and the need to build on more difficult soil conditions (not to mention a shortage of suitable trees in some areas!) created a need for alternative approaches, and deep foundations evolved through the early hand-dug concrete caisson and the driven steel pile to the current range of alternatives – drilled shafts, displacement shafts, pre-cast concrete piles, steel piles of various configurations and augered, cast-in-place piles.

The history of nondestructive test methods for deep foundations is almost as hard to pin down as the history of deep foundations themselves, but one thing is certain – development of NDT occurred along parallel paths in several different parts of the world. The present state-of-the art is a result of knowledge and experiences shared at international conferences sponsored by such groups as the Deep Foundations Institute (DFI International), the International Association for Foundation Drilling (ADSC-IAFD), the International Society for Soil Mechanics and Geotechnical Engineering (ISSMGE), and in research projects sponsored by professional or government bodies, such as the Federal Highways Administration (FHwA) in the USA, and various European Departments of Transportation or construction industry research centers, such as the Centre Experimentale de Recherche et d'Études du Bâtiment et des Travaux Publics (CEBTP) in France, The Netherlands Organization (TNO) in Holland and the Federal Institute for Materials Research (BAM), Berlin, Germany.

The scientific principles behind some modern test techniques can be traced back to Victorian times. A graduate student at Northwestern University, Illinois, USA, while completing a literature search for his Ph.D. dissertation on the 'Frequency Equation for Cylindrical Piles Embedded in Soil' (Hannifah, 1999), found two 19th Century wave-propagation research references, one of which dates back to 1876 (see Chapter 13 – Current Research: Guided Waves).

By the mid-1950s, it was well-established that the propagation velocity of a stress wave through concrete was a function of the modulus and density of the material, and researchers had begun to look at ways of using stress waves to assess the quality and integrity of deep foundation shafts. While the theory seemed simple enough, electronic technology lagged far behind the researchers' needs, and stress-wave measurements proved very difficult to put into practice outside of the laboratory.

The first published reference to measurement of high-strain stress waves in a driven pile was made in England in 1938 (Glanville *et al.*, 1938), but it was more than 20 years before practical applications for high-strain stress-wave measurements were developed. The breakthrough research was conducted more or less concurrently by teams at Case Institute of Technology (now Case Western Reserve University) in the United States, and the building research division of the Dutch Technical Research Institute, The Netherlands Organization (TNO) in Holland.

In Europe, the Dutch first discussed high-strain measurements in TNO's in-house publication 'TNO Rapport' (Report). TNO Report Number 341, published in 1956, described stress-wave measurements during the driving of three piles for Jetty No. 1 in the Rotterdam Harbor project (Verduin, 1956). At about the same time, an article which discussed what occurred in the soil during pile driving was published by De Josselin De Jong in *De Ingenieur*, a Dutch-language engineering publication (De Josselin De Jong, 1956). Henk Van Koten led much of the TNO research at that time, and published a paper on stress-wave propagation in a driven pile in 1967 (Van Koten, 1967).

According to the recollections of the Case team, divulged in personal correspondence with the authors during the preparation of this book, the development of high-strain stress-wave testing in the USA began with a 1958 Master's Thesis by a student named Eiber at Case Institute of Technology, in Cleveland, Ohio. This thesis led to the establishment of an extensive research project under the direction of Dr George Goble with funding by the Ohio Department of Transportation and the United States Federal Highway Administration (FHwA). Started in 1964, the Case team's research determined that both strain and acceleration measurements at the pile top were necessary for dynamic pile analysis. Early measurements were recorded on oscillographs, which used tiny mirrors mounted on electrical armatures that were driven by the input signal, to reflect a narrow beam of light onto photosensitive paper. By 1970, high-accuracy magnetic tape recorders were available to record the data.

The real-time analysis, termed the 'Case–Goble Method', is named after the University and the Research Director (Goble, 1967; Goble *et al.*, 1975), while the more extensive numerical analysis CAPWAP (Case Pile Wave Analysis Program) is a modeling technique that uses the high-strain measurement data to determine the dynamic response of the soil (Rausche *et al.*, 1972).

The Case researchers were limited by the technology of the day, just as the TNO team was, and practical equipment became commercially available at about the same time in both Europe and the United States. Goble and his associates from Case formed Pile Dynamics Incorporated and developed their first commercial equipment in 1972. The earliest equipment for on-site recording and analysis relied on analog computers. Digital computers became generally available both in Europe and the United States in the early 1980s. Current equipment for NDT of foundations is based on either PCs or hand-held computers, and in some cases data can be transmitted via modem or cellular telephone from the construction site to the engineer's office in a matter of minutes after completion of testing. This progression of capability reflects the growth in electronics technology over the past two decades.

Although high-strain dynamic pile testing was developed initially to determine bearing capacity and/or hammer efficiency, it was quickly realized that evaluation of driving stresses and identification of pile damage also provided valuable information. These features were soon incorporated as a standard part of the pile-driving analysis procedure. In 1974, researchers began to consider application of these techniques to drilled shafts.

The low-strain Impulse-Echo (or Sonic-Echo) test was developed in the 1960s. One of the leading researchers to explore the capabilities and limitations of the Impulse-Echo method was Jean Paquet, of the Centre Experimentale de Recherche et d'Études du Batiment et des Travaux Publics (CEBTP) in St. Rémy-les-Chevreuses, France. The CEBTP is the research institute established by the federation of the construction industry in France and is primarily concerned with construction quality control. Paquet was a prodigious researcher who simultaneously directed research and development programs concerning high-strain, low-strain and ultrasonic testing of deep foundations, plus allied programs concerned with developing the software for analysis of the data from these methods. The technology of the time made the Impulse-Echo method difficult to apply in the field, and it was not commercially available until 1974 in Europe and 1979 in the USA. In both cases, the low-strain Impulse-Echo test was a derivative of the high-strain pile-driver analysis technique.

The earliest versions of the Impulse-Echo test used a hammer impact to generate a stress wave, and an oscillograph, or UV recorder, to record the response of a geophone or an accelerometer attached to the top of the shaft – a painstaking process that required delicate timing and often resulted in a lot of wasted paper. The first major improvement came in the early 1970s, when the phosphor storage oscilloscope replaced the oscillograph. A photograph was taken of the test result on the oscilloscope screen, which still required careful timing and often resulted in blurry images and much wasted film! The next major advance was the advent of the microprocessor and in 1984 TNO researchers announced the first digital version of the Impulse-Echo test (Reiding *et al.*, 1984, Schaap and de Vos, 1984). The USA team followed with its own version in 1985.

The Impulse-Echo test was developed for use on pre-cast, driven piles, and works very well in reasonably soft, uniform soil conditions where shaft length and cross-section are known. However, it is often less conclusive on sites with highly variable soil conditions or on drilled shafts where shaft cross-section can vary due to use of temporary casings or variations in lateral soil stiffness. The effective penetration depth of the method is also limited by the stiffness of the lateral soils.

As the Impulse-Echo test was evolving, other visionaries in the United States also saw the potential for NDT of drilled shafts. One of the more notable was the late John P. Gnaedinger, founder of STS Consultants, Ltd. Gnaedinger developed and patented the G-cell, which consisted of a small steel cell which contained a remote-controlled striker assembly, similar to that in a door chime. The G-cell was installed at the base of a shaft attached to the reinforcing cage. After concrete had been placed and reached a reasonable maturity, a sensitive sound recorder was attached to the top of the shaft and the G-cell striker was activated. An oscilloscope measured the time taken for the stress wave generated by the striker to reach the top of the shaft. Since the stress wave was only traveling one way, instead of down the shaft and then back up again, the G-cell method could potentially be used on shafts twice as deep as those that could be tested by the Impulse-Echo method. Gnaedinger's US patent, 'US 3 641 811: Method and Apparatus for Determining Structural Characteristics', was filed in 1969 and granted in 1972 (Gnaedinger, 1972). No fewer than 14 subsequent applications for nondestructive test patents referenced Gnaedinger's patent.

In 1968, Paquet published a landmark paper that discussed the limitations of the Impulse-Echo method and described the assessment of drilled shafts by the 'Vibration method', in which a swept-frequency vibrator was attached to the head of the shaft and the response monitored by multiple velocity transducers (calibrated geophones) (Paquet, 1968). A major drawback of the method was the amount of preparation required to attach the vibrator and the transducers to the top of the shaft. Paquet was visionary enough to understand the difficulties of applying the Vibration method reliably under actual site conditions and to foresee the development of the Impulse-Response method, in which a hammer blow through a calibrated load-cell would generate a quantified impulse, and thus allow a network analysis of the shaft response to a known input, even though the electronics at the time were not capable of recording such an event.

In 1974, Paquet applied for a patent on an analysis method that used a Fast Fourier Transform (FFT) of the recorded data into the frequency domain, where velocity was divided by force to provide the transfer function, or mobility signature, of the shaft. It was 1977 before analog computers finally caught up with Paquet, and technicians could make his theory become a reality. Since then, the development of NDT methods has proceeded rapidly. The introduction of digital computers has revolutionized the field, and the Impulse-Response method is now widely accepted throughout the world for the assessment of drilled shafts, locating voids beneath pavement slabs and behind tunnel linings and assessing concrete quality in structures ranging from parking decks to chimneys and storage silos.

The stories of the Cross-Hole Sonic Log (CSL) and Parallel-Seismic methods are similar. These methods were also developed by the CEBTP in the late 1960s (Paquet, 1969; Paquet and Briard, 1976) but were hampered by the technology of the time. The advent of the portable digital computer, and then the PC, made CSL testing an inexpensive reality and it began to be widely used in Europe in the early 1980s. Several countries in Asia and North Africa quickly followed the European lead, largely as a result of French post-colonial influence, but the CSL method was not introduced commercially to the United States until 1986.

Despite widespread use of NDT methods for both driven and drilled shafts in Europe by the early 1980s, NDT for drilled shafts was much slower to be adopted in the United States, Canada and South America. The first use of the CSL method in the Americas was by the present authors in 1986. Drilled shafts were constructed in the Spokane River for the repair of the flood-damaged powerhouse at the Upstream Dam Hydroelectric Project in Spokane, Washington. The construction conditions were extremely difficult because the river was still in flood, and so the owner decided that it would be a good time to try CSL to verify the quality of the foundations. The test immediately proved its value in conditions where no other form of testing was practical.

In 1988, the FHwA and the California Department of Transportation (CALTRANS) sponsored a research program in which a number of drilled shafts were constructed with known defects, and NDT practitioners were invited to test the shafts with whichever methods they chose. The interest created by the project was considerable, and it was extended to include additional shafts installed on a site at Texas A&M University in College Station, Texas. The end result, published in the FHwA Report No. FHWA-RD-92-004 (Baker *et al.*, 1993) was one of the most critical factors in the acceptance of NDT for drilled shafts in the United States. The FHwA report started reaching the desks of State DOT engineers and specifiers late in 1994, and they began specifying NDT methods for quality control by about the summer of 1995. In 1996, the general construction industry started to follow suit, and by the year 2000 the use of NDT for both driven and drilled shafts was almost commonplace in the USA.

The effect of this acceptance of NDT methods went beyond the testing community in most countries. The fact that there is now a number of quality assurance techniques available for drilled shafts, where previously there had been no economically practical method of testing the finished product, has encouraged many engineers to use drilled shafts for projects where they would formerly have preferred to use driven piles. The same has occurred with the use of drilling slurry. Now that the end-product can be closely examined, engineers are more comfortable in allowing excavation and concrete placement under water or slurry, instead of requiring the foundation contractor to adopt expensive multi-casing methods in an effort to seal out groundwater and unstable soils so that concrete can be placed 'in the dry'.

Unthinking reliance on the FHwA report or other similar references, however, can have the effect of 'freezing' the state of the technology. As mentioned in the Preface, the need for this book was illustrated recently by a project in the United States in which the engineer was relying on the FHwA publications. A problem had occurred during the construction of a drilled shaft, which gave rise to the possibility of soil contamination of the concrete about halfway down the shaft. The shaft was not equipped with access tubes for any of the down-hole tests, and so the testing firm for the project recommended a surface reflection technique such as Impulse-Echo testing. The project engineer refused to accept that proposal because the FHwA report gave the impression that surface acoustic methods were only capable of detecting a major defect in a drilled shaft, and the subsequent FHwA publication 'Drilled Shafts: Construction Procedures and Design Methods' (FHwA-IF99-025) contains the recommendation that these methods should not be used as the primary integrity testing method for axially loaded shafts in which the design load exceeds 40 % of the structural capacity. This may well have been the case in 1993, but research and development of test equipment and analysis procedures has continued unabated since then. The ability of the surface reflection techniques to detect smaller anomalies is governed by several factors, including the length/diameter ratio of the shaft, the depth to the anomaly and the type of anomaly. In these authors' experience, anomalies as small as 10 % of the shaft cross-section can sometimes be detected by surface reflection techniques. By blindly sticking to the opinions expressed in the FHwA report, the engineer was refusing to acknowledge several important facts:

- The report was based on data gathered more than ten years ago, from about 1988 to 1991.
- Some of the personnel involved had less than three years experience with the techniques at that time.
- The hardware and signal quality has improved significantly since then.
- Research has continued into better data acquisition and analysis algorithms.

In a more recent Class-A prediction study at the National Geotechnical Experimental Site (NGES) at the University of Massachusetts in Amherst, MA, it was concluded that, in fact, the skill and experience of the operators, coupled with more advanced equipment, had improved the overall performance of surface NDT techniques to the extent that some of the better participants located defects as small as 6 % of the shaft's cross-sectional area, and multiple defects were accurately located in some shafts. It was conceded, however, that the skill of the operator was crucial to the success of the surface NDT methods (Iskander *et al.*, 2001).

In these authors' experience, and the experiences of other reputable experts in the field of nondestructive testing, the smallest anomaly that can be reliably detected and quantified by the surface reflection techniques is about 10 to 15 % of the shaft's cross-sectional area, depending on shaft dimensions, anomaly depth and soil properties. Whether such an anomaly is significant or not must be judged on a case-by-case basis by an experienced engineer, rather than by the blanket rejection implicit in the 1993 FHwA report.

Most contractors and engineers will readily admit that the feedback provided by NDT has enabled them to refine shaft designs and construction techniques, and modify equipment to improve the quality and reliability of drilled shafts and augered, cast-in-place (ACIP) piles. This, in turn, has increased confidence in drilled shaft and ACIP foundations, and made a significant contribution to the growth of their respective market sectors.

1.3 DEEP FOUNDATION FAILURES AND NDT

Well-documented failures of the shafts of piled or drilled shaft deep foundations in service are rare. Two possible reasons for this are the tendency to 'over-design' deep foundations, thereby reducing risk of failure to a minimum, and difficulty in distinguishing between failure of the shafts and failure of the soil bearing capacity, or indeed, a combination of both. However, the rare reported failures are a warning to the underground industry, highlighting the difficulties in predicting the variables and unknowns present, particularly in water-bearing soils. As the FHwA report (Baker *et al.*, 1993) states, 'this results in a lower risk tolerance for a single or double shaft supported pier compared to multiple piled foundations'. The modern trend indeed is to replace the latter with large-diameter drilled shafts, particularly for large bridge structures in seismic zones.

Some settlement of foundations is expected, and allowed for in the design of the structure. Determining exactly what amount of settlement should be considered unacceptable is problematic, and the amount of settlement that constitutes failure of the foundation is even more contentious. Lessons learned from shaft failures do offer help in avoiding problems in future construction; however, a complete, impartial investigation of deep foundation failure is costly, and therefore rare. Any investigation is usually funded by legal costs alone, and litigation does not necessarily produce a clear picture of the whole story. The first two examples described here are typical of this dilemma.

1.3.1 ESSO OIL TANKS, FAWLEY, HANTS, UK

An oil refinery tank farm was constructed on soft soil conditions in Southern England to accommodate the large quantities of oil being delivered by the supertankers that were built following the closure of the Suez Canal. Compared to existing design practice, the size of tanks required to store the oil arriving at the refinery increased dramatically. One tank failed during water test loading and a second tank showed incipient foundation failure during the start of water test loading. The full story is described in Leggatt and Bratchell (1973) and the Institution of Civil Engineers (UK) (1974) gives a brief description of the events. Driven cast-in-place 420-mm-diameter piles with expanded bases were founded in river gravel through approximately 2 m of gravel fill overlying 6–8 m of soft silty clay and peat. The piles were reinforced to the base, and the casing was withdrawn during concreting. A thickened portion of the 380-mm-thick reinforced concrete raft tank base capped each pile. After tank failure, a tunnel was driven beneath the raft to expose the upper part of some piles. At the location of maximum raft deflection, many piles showed signs of bending overstress, increasing in severity from vertical cracking to complete separation of the pile from the pile cap. In addition, some shafts had visible necking, with exposure of reinforcing steel. Nondestructive vibration testing was performed on 43 of the exposed shafts.

Experts were appointed to examine the failure, representing the owner on one hand and the piling contractor on the other. The experts for the owner claimed defective shaft construction as the cause of tank failure, while the experts for the piling contractor claimed failure of the founding gravel layer as the reason. The case was settled before the end of legal proceedings, with no clear technical agreement on the root cause of the failure.

1.3.2 NEUMAIER HALL, MOORHEAD, MN, USA

The second example concerns a fifteen-story residence hall built in 1969 on an American university campus. The following case history emerged from research on the Internet. The building was constructed using lift-slab architecture, with reinforced concrete floor slabs carried by steel columns. These columns were in turn supported by concrete caissons (drilled shafts) extending 30 m down through the clay subsoil, terminating in bell-shaped bases drilled into dense glacial till below. After construction, the north-west corner developed large cracks in the foundation and on interior and exterior walls. These cracks were the result of differential settlement of the foundation and partial rotation of the building, which exceeded 75 mm in certain areas. The settlement was believed to be induced by overstress of the caissons, possibly due to negative skin friction as a result of a lower water table, loss of integrity of the caissons due to necking or contamination of the concrete and a complete structural failure of the caisson shaft. One of the unique observations is that the caisson in the north-west corner, despite complications during its construction, had been seemingly immune to these effects, and had settled within the expected limits. Due to concern that the strain on structural components could lead to failure of the lift-slab architecture, and an estimated repair cost of over one million dollars, the decision was made to close the building and demolish it.

Apparently, the building first began to show signs of foundation problems within ten years of its construction, with foundation columns showing a differential settlement of 25 to 30 mm. Measurements of the site were taken to observe the settling rates, but were complicated by apparent movement of other parts of the structure. Once a stable benchmark was established, all readings began showing a general negative (downward) movement. The continuing stress on the structure as the foundation settled soon became evident. Major cracks and displacements were visible on the north-west corner of the building over the first three stories. Interior foundation walls showed cracks, again in the north-west corner. Several rooms in this corner were rendered unusable because of the extensive cracking; also, windows and doors became non-functional due to the stress. At the fifteenth floor, large horizontal cracks appeared all along the north wall. A maximum differential settlement of nearly 100 mm and a total settlement of more than 125 mm were reached. In addition, much of the stress represented itself in the form of rotation. The entire structure was attempting to rotate about the point of greatest stress. Measurements showed that, in places, the building had deviated from plumb by more than 75 mm. By comparison, the values generated when evaluating the original design load indicated that the expected settlement was in the area of 25 to 33 mm. It is interesting to note that the north-west shaft settled a total of 40 mm, close to the expected amount. A list of possible 'culprits' was created:

- Settlement of the hard glacial till (hardpan).
- Settlement of the clay sands above 30 m.
- Settlement of disturbed material at the bottom of the caisson bells, caused at the time of construction.
- Structural compromise of caissons.

The first to be considered was hardpan settlement. The hardpan is composed of glacial sandy–gravelly till. The recorded standard penetration N value for this material under the structure was 50. For that N value, a settlement of 125 mm is not a realistic probability.

The second consideration was settlement of the clay sands above 30 m. The elevation of the bases of the shaft bells was also brought into question. They must be located on the correct bearing stratum in order to support the structure. If the shafts had not been drilled deep enough, that could perhaps account for the settlement. The glacial till had been designated as the bearing stratum for Neumaier Hall, and a study of the numerous geotechnical test borings performed on the site showed that the bearing strata was encountered at about 30 m depth. Shaft construction records showed that the shafts were founded at or below this level.

A third possibility was settlement of disturbed material at the bottom of the shaft bells, caused at the time of construction. It could not be determined from the construction records or the inspector's reports if there actually was disturbed material at the bottom of the bells. Calculations, however, showed that any settlement caused by disturbed material at the base of the shafts would have taken place rapidly, and not exhibited the long-term behavior that was recorded on this site.

The fourth consideration was structural inadequacy of the drilled shafts. It is quite possible that several unrelated deficiencies, none of which would be significant by itself, could occur in the same shaft, where the combined effect would result in a significant reduction in shaft capacity. The following scenarios have been well documented and could combine to cause failure of the shaft:

- If the temporary casing is removed too rapidly, the concrete within the casing can 'arch' or lock itself into the casing, thus causing it to be lifted with the casing, which creates suction that may draw soil and water into the concrete, hence resulting in a reduction in cross-section, or 'neck-in'.
- Similarly, if the temporary casing is removed too late, after the concrete has begun to set, the same scenario can occur – some or all of the concrete is lifted with the casing, thus creating a discontinuity in the shaft and a suction that draws soil and water into the resulting void(s) within the shaft.
- The handling and mixing process, particularly if water is added at the site to make the concrete more workable, can affect concrete strength.

A fifth possibility was negative skin friction on the shaft caused by consolidation of the surrounding soils. A significant reduction in the ground water table was recorded on the site, hence causing considerable consolidation of the soil. One of the engineering companies that investigated the failure later calculated that the down-drag or negative skin friction caused by the soil consolidation could have caused the loads on some shafts to be 170 % of the designed capacity. The original design for the caissons anticipated a load of 0.7 MPa. The total down-drag loads plus structural loads were calculated to be between 0.97 and 1.38 MPa at the bottom of the shaft bells. Since a load of more than 2 MPa would have been necessary to cause the 75–125 mm settlement in the glacial till, it is highly probable that a reduction in either shaft diameter or concrete quality occurred, which raised the stresses in that part of the shaft to the level required to cause failure.

1.3.3 TAMPA CROSSTOWN EXPRESSWAY, TAMPA, FL, USA

Some widely reported foundations failures are clearly proven to be failures of the supporting soil and therefore a result of inadequate site investigation prior to designing and constructing the foundations, but the distinction is rarely made in the public media. A very recent example is the settlement of highway viaduct foundations in Tampa, Florida, USA.

An elevated reversible-lane tollway built was being built in the median of the Lee Roy Selmon Crosstown Expressway. When the launching gantry for the deck segments was being positioned on Pier No. 97 on April 13, 2004, the pier sank 11 ft into the ground. A subsequent investigation found that the soil conditions at the pier location were inadequately defined before the design was completed. News media reported that the normal procedures for the local area had been followed, which included drilling an exploratory borehole at each pier location. Unfortunately, the borehole at Pier No. 97 apparently encountered only a limestone pinnacle or ledge, rather than a solid bed. Once the load of the launching frame settled on the pier, it forced the pier to 'punch through' the limestone into the surrounding soft sediments. Two 150-ft sections of roadway buckled as a result.

A few months later, on July 6, 2004, Pier No. 99 settled 1.3 in, which was beyond the acceptable limit of 1.0 in. While an official report on the subsequent investigation has not been released at the time of writing this manual, local news media reported that, of 215 piers, excessive settlement was believed to have occurred on a total of 154.

1.3.4 YUEN CHAU KOK, SHATIN AREA 14B, PHASE 2, HONG KONG

Regrettably, it is not just accidental omission or misfortune that causes problems with deep foundations. It seems to be a universal truth that foundation contractors are among the most 'at-risk' contractors on a construction project, largely because they are dealing with subterranean conditions that are often poorly documented and full of surprises. That fact often combines with the widespread propensity for accepting the lowest bid as the most appropriate bid, regardless of qualifications, to leave the foundation contractor struggling against unforeseen conditions with a minimal budget. This sometimes puts foundation contractors in such a financial bind that one or two of them resort to 'fraud' in order to get paid. Such a case was recently publicized by the Hong Kong Housing Authority (HA) on its Internet website (Hong Kong Housing Authority, 2000).

The report publicizes the findings of a panel that was convened to investigate the circumstances that caused excessive foundation settlement and led to the forced demolition of two partly completed buildings of more than thirty stories. According to the HA report, a foundation contractor that was prequalified with HA won a contract for constructing large-diameter bored piles and installing driven piles on the Yuen Chau Kok, Shatin Area 14B housing project, Phase 2. The project involved the construction of deep foundations for two forty one-story apartment blocks, three thirty three-story blocks and a car park/ancillary facilities building. The large-diameter foundation shafts for the tower blocks were designed to be founded on competent bedrock, with appropriate under-reams, or 'bells'. Unknown to the HA, the winning foundation contractor subcontracted the large-diameter drilled shafts to another contractor who apparently did not meet the requirements for prequalification with HA.

Alerted by excessive settlements that had been reported on foundations on other HA projects, the HA decided to monitor the settlement of foundations on all sites, including the Shatin project. Excessive settlement was recorded at Shatin as the tower block superstructures evolved, becoming so severe that the project was halted when the towers had reached about thirty stories. The settlement continued, and the tower blocks were eventually deemed unsafe, and demolished. Forensic examination of the foundations by full-depth core-drilling showed that, out of thirty six shafts on the site, fifteen were founded short of bedrock by up to 1 m, ten were between 1 and 5 m short of bed rock, and eleven were between 5 and 15.4 m short! Only four of the shafts were proven to be founded on the bedrock as designed, and only eight were composed of concrete that met the project's quality requirements. The other shafts showed evidence of 'honeycomb' concrete, steep or vertical jointing and fractures.

After examining concrete volumes and steel lengths delivered to the site by reputable firms with robust quality control systems, the Enquiry Panel considered the difficulty of disposing of or re-routing substantial quantities of steel and concrete without attracting attention. The Panel also concluded that the foundation contractor had most likely drilled the full depth of most shafts, but did not use temporary casing over the full length, as required in the specification. The Panel also concluded that the sidewalls of the shafts collapsed early in the concrete placement process, and the concrete simply filled the voids created by the sloughing soil – thus, the shafts ended up with substantial 'neck-ins' or total discontinuities.

The Enquiry Panel found compelling evidence of fraud on the part of the foundation contractors. The Panel noted that inspection of the shafts by Cross-Hole Sonic Logging (CSL – see Chapter 10 of this book) had been specified, but the majority of the CSL access tubes had been deliberately 'blocked' to hide the fact that the shafts either contained significant defects or were shorter than the designed length. As a substitute for the inconclusive CSL testing, the foundation contractors offered vibration test data instead (see Chapter 8 – Impulse-Response testing). Unfortunately, the responsible engineer for the HA was unaware of the limitations of the Vibration test and was also unaware that most of the test data were in fact from an adjacent structure, fraudulently presented as being from the structures in question. Finally, when core samples were demanded from the shafts, it is apparent that the HA made the mistake of letting the foundation contractors engage and supervise the core drillers. The Enquiry Panel determined that, in several cases, multiple shallow cores were drilled and then 'doctored' by the foundation contractors to form a composite core that was presented as being from the full length of the shaft in question.

The Enquiry Panel found that the contractors had deliberately obstructed the performance of the tests, manipulated the testing companies and faked both physical and nondestructive test results in order to gain approval for the shafts in question. The enquiry revealed that the subterfuge and collusion had gone unnoticed for so long because the foundation contractors repeatedly worked late into the night, long after the HA inspectors had left the site for the day.

Lack of adequate inspection obviously played a significant role in what could have turned out to be a major fatal disaster had the excessive settlement not been noticed early in the project. It is stated in the HA Enquiry report that there had been too much bureaucratic reliance on correct paperwork, and not enough actual on-site inspection. The reliance on properly completed paperwork generated a false sense that quality assurance was well under control. Similarly, it is implicit in the report that most of the faked NDT data would have been discovered much sooner had the inspectors and engineers in question been more conversant with both the construction technique and the capabilities and limitations of the NDT methods used.

Fortunately for us all, most foundation contractors are conscientious and skilled specialists, but the salutary lesson to be learned from the HA experience is that nothing can be taken for granted, even though all of the paperwork seems to be in order. No amount of paperwork can substitute for conscientious and experienced inspectors, supported by a carefully designed program of nondestructive testing performed by competent field personnel and analyzed by experienced specialists.

1.4 DEFICIENCIES IN EXISTING FOUNDATIONS

Integrity testing of deep foundations as a quality control tool is intended primarily to reduce the number of defective shafts and does not address the geotechnical behavior of the foundation, although the Sonic-Mobility test does offer some insight into soil conditions by the measurement of dynamic stiffness. These present authors have tested foundations beneath demolished structures some years after their construction. Shaft defects discovered at this time in several cases were attributed to built-in deficiencies at the time of construction and not at demolition. Three examples stand out in this group.

The first is a chemical refinery destroyed in an explosion that removed all above-ground structures, leaving the foundation pile cap bases intact. As part of an attempt to assess the viability of rebuilding the plant, the piled foundations were tested non-destructively through the reinforced concrete bases. Each pile cap had three to four 450-mm-diameter driven cast-in-place piles, and the NDT failed to locate 20 % of the piles through the bases. NDT inadequacy was suspected and the bases with no pile response were excavated to reveal that the concrete in the shafts at their junctions with the pile caps was partially or totally missing! These defects had not stopped the structures operating successfully for at least twelve years.

The second example, a multistory parking garage demolished forteen years after construction to make way for a new high-rise office building, affords a second example of defective shafts beneath structures previously in service. The foundation system comprised evenly spaced 900-mm-diameter bored piles (drilled shafts) linked by ground beams. It was hoped that most of these piles could be incorporated in the new structure design and their heads were exposed for nondestructive testing. The authors performed NDT on the shafts to evaluate their suitability for re-use. A number of these piles showed considerable necking in the upper 3 m of their shafts and were excavated for visual assessment. Some shafts had no concrete for at least 50 % of their cross-section, over lengths up to 2.5 m. Again, the parking garage had performed with no problems for forteen years.

The third example is a utility chimney that collapsed during liner cleaning work. The cause of the collapse was believed to be a large mass of flyash that became dislodged all at once, falling several hundred feet to the bottom of the liner, where its kinetic energy was deflected radially outward in an explosive manner, effectively cutting the chimney away from its foundation. Since the foundation slab appeared to be relatively undamaged, the present authors were called in to evaluate the piles supporting the slab to determine their suitability for re-use if a new chimney was to be built on the existing foundation. Once again, several shafts could not be detected by impulse testing from the surface and were physically investigated by excavation along the edges of the foundation slab. In three cases that were investigated, the corrugated steel permanent casing was cut open and the joint between the top of the pile and the underside of the slab was found to consist of unbonded gravel to a depth of several inches. Prior to the structural failure of the chimney, the foundation had shown no signs of excessive or differential settlement.

These cases support the contention that there is considerable 'over-design' in many deep foundations. However, they should not be interpreted to mean that it is not important to control the quality of deep foundations, both from the viewpoint of cross-sectional integrity and material quality. These examples also raise the question of what constitutes an unacceptable defect. Joram Amir, of PileTest in Israel, has made several presentations at Deep Foundations Institute seminars and other geotechnical engineering venues (e.g. Amir, 2002) in which he recommended that anomalies identified by nondestructive test methods should be classified according to severity:

- Anomaly – this could be just an anomaly in the test data caused by the equipment (noise, etc.), the means of access (e.g. access tube debonding) or site circumstances (electrical interference, noise, etc.).
- Flaw – an imperfection or irregularity of shape or material, but not significant in terms of shaft capacity or durability.
- Defect – affects the bearing capacity or the likely durability of the shaft. Engineering evaluation is required – perhaps the shaft can be accepted at a reduced capacity?

The present authors agree with Dr Amir's recommendation. It should be the testing company's responsibility to evaluate the first possibility. Before attempting any analysis, the testing company should verify that the anomaly was not caused by an equipment malfunction, operator error, site circumstances or data-storage problem. Depending on the test method in use, this may be as simple as repeating the test two or three times and then comparing the results. Once equipment or application anomalies have been eliminated, only then can the anomaly be considered a real flaw in the shaft and investigated further to assess its nature, location and significance. Techniques, such as tomography, core drilling or excavation, may be required to provide definitive information. If the flaw is proven to exist, the geotechnical and structural engineers together should assess its significance, taking into consideration its nature, size and location within the shaft. If it is considered likely to cause an unacceptable reduction in shaft capacity and/or durability, then it should be considered a defect and therefore a reason to reject the shaft until it has been remedied.

Universal acceptance criteria for situations such as these would go a long way towards standardizing test methods, analysis of test data and reporting procedures. At the time of writing this book, the Testing and Evaluation Committee of the Deep Foundations Institute is in the process of creating a document that could include a general consensus guideline for drilled shaft acceptance criteria based on the results of Cross-Hole Sonic-Logging tests and a sample specification for engineers and owners to use (see Chapter 10 of this book for more details).

2

Commonly Used Deep Foundation Construction Methods

It is important that anyone who is to specify or perform nondestructive testing of deep foundations has at least a basic familiarity with the various methods used to construct the foundations. The likelihood of a defect, the probable nature and cause of it and its likely significance to the shaft's capacity are all variables which change according to the foundation construction method used. All too often the present authors have been called to peer review or give a second opinion in cases where the testing personnel had no knowledge of foundation construction methods and either chose inappropriate procedures for the site in question or issued inconclusive or ambiguous reports. This chapter provides a very brief overview of the various construction methods, partly to assist the novice reader in understanding the complexity of the deep foundation business and hopefully to spark enough interest in the topic that the interested reader will seek additional knowledge.

There are four basic deep foundation construction methods but many variations of the four basic methods are commonly used. In some cases, the variations are proprietary and patented, but have proven so successful that yet further variants are developed as competitors endeavor to create the same product without infringing the patents. All modern deep foundation construction methods are based one of the following techniques:

- Driven Pile – a prefabricated timber, steel or concrete pile is driven into the ground by impact or vibration.
- Driven Cast-in-Place Pile – a hollow steel casing is driven into the ground, filled with the foundation material and then removed.
- Drilled Shaft – a hole is excavated in the ground and filled with the foundation material.

Nondestructive Testing of Deep Foundations B.H. Hertlein and A.G. Davis

- Augered, Cast-In-Place Pile – an auger is screwed into the ground and the foundation material is placed through it as the auger is withdrawn.

Each foundation type and proprietary variation presents its own unique set of challenges to both the contractor and to the inspector charged with documenting the process for quality control and payment purposes.

2.1 DRIVEN PILES – TIMBER, STEEL AND CONCRETE

Driven piles, as their name implies, are driven into the ground by mechanical force. The definition used to be simply that the pile was banged into the ground by a hammer dropping under the force of gravity but things have now changed. Today, driven piles can indeed be banged into the ground by either an air- or a diesel-powered hammer but they can also be vibrated into place by a hydraulic shaker that effectively liquefies the soil. In certain soil conditions, if noise or vibration is unacceptable, piles can even be pushed into place by sheer brute force.

Timber piles, the original and still, in some cases, the most economical driven foundation, often surprise engineers with their simplicity and durability. Smith's article in the DFI's *Deep Foundations* magazine (Smith, 2005) refers to a previous publication by Fleming *et al.* (1985) which describes the exposure of timber piles beneath a bridge over the Danube that had been built by the Roman Emperor Trajan in 104 AD. When exposed during the 18th Century, the piles were found to be 'petrified to a depth of 20 mm (0.75 in) and that the timber beneath this surface layer was completely sound'. During redevelopment projects in London and Liverpool, UK and Chicago, USA, the authors tested a number of timber piles that were more than 100 years old, and found them to be in almost as good condition as when they were installed. The testing included NDT, trial pits for visual examination, mini-core sampling and, in the case of Tobacco Dock in London, actual extraction of the piles after NDT. The test data confirmed the often-quoted theory that rotting of timber is an aerobic process and is unlikely to occur in an anaerobic environment, such as below a stagnant water table or deep in permanently saturated soils.

Timber was also used to make sheet piles. The latter are often used to support temporary soil excavations or to form 'cofferdams' in rivers, lakes and marine environments. The water inside the cofferdam is pumped out, enabling construction work to be performed below normal water level under almost dry conditions. The crudest sheet piles were just adjacent planks hammered into the ground, usually tied together near the top of the excavation by a crossbeam or 'waler'. Where 'water-tightness' was a necessity, however, better inter-sheet joints were needed. One of the earliest records of timber sheet piles with an engineered joint is a patent granted to Wakefield in 1887, which describes three timber planks bolted together with the center plank offset to form a tongue-and-groove structure (Figure 2.1(a)). Alas, despite extensive searching via the Internet, the present authors were unable to find out much more

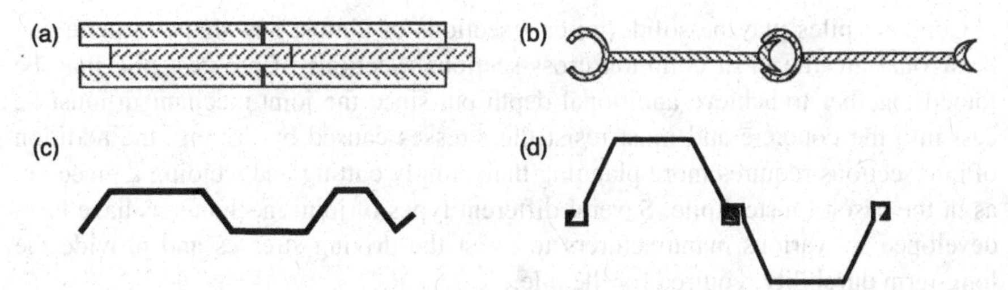

Figure 2.1 (a) Cross-section of Wakefield timber sheet piles (circa 1887). (b) Cross-section of early steel sheet pile (circa 1905). (c) Modern lap-joint steel sheet pile. (d) Modern 'Larssen' section interlocking sheet piles

about Wakefield, the inventor. The Wakefield sheet pile, however, is well-known, and mentioned in documents published by the California Department of Transportation (CALTRANS), and on several Internet websites concerned with sheet piles. The basic tongue-and-groove concept behind the Wakefield sheet is still in use today. If you would like more information, try your own search using the keywords 'Wakefield sheet pile'.

Steel piles may be 'H-section' beams, cylindrical tubes or sheets bent into various profiles. If greater length is required in steel piles, additional sections are simply welded on as the existing sections are driven close to ground level. Steel sheet piles used for cofferdams typically have some form of connector on each vertical edge to link to the adjacent sheets and thus form a continuous wall with locked joints (Figure 2.1(b)), unlike the Wakefield timber sheet, which had only a tongue-and-groove joint. A slightly different variation of Wakefield's simple over-lapping concept is the modern standard lap-joint steel sheet (Figure 2.1(c)).

For many years, there were only two or three manufacturers of steel sheet piles, offering about seven or eight cross-section shapes between them. Today, there are at least ten manufacturers around the world, offering a total of more than 200 different combinations of cross-section and interlock mechanism between them. For the reader that is particularly interested in sheet piles, an Internet search using the keywords 'steel sheet piles' will return links to several very informative and useful sites.

Displacement piles are a variation of the tubular pile, where a casing is driven into the ground, displacing the soil laterally. A reinforcing cage is placed and then concrete is poured inside the steel casing. Sometimes, the casing is left in the ground as a permanent part of the pile but more commonly the casing is removed before the concrete sets. In one type of displacement pile, developed in the 1920s by British Steel Piling (BSP), the casing was extracted by reversing the action of the hammer so that it struck in an upward direction against haulage links that were attached to the casing. This tapping action helped consolidate the concrete without the need for additional vibration. This construction method was patented as the 'Vibropile'. Another variant of the steel displacement pile is the tapered pile which, just as its name suggests, tapers in section, being narrower at the bottom, or 'toe', than at the top.

Concrete piles may be solid, pre-cast sections of almost any shape. Square and hexagonal are the most common cross-sections. Sections of pre-cast pile may be joined together to achieve additional depth but since the joint mechanism must be cast into the concrete and must resist the stresses caused by driving, the addition of pile sections requires more planning than simply cutting and welding a piece on, as in the case of a steel pile. Several different types of joint mechanisms have been developed by various manufacturers to resist the driving stresses and provide the long-term durability required for the pile.

Cylindrical concrete 'shell piles' come in a variety of forms. The spun concrete pile is a very dense reinforced concrete tube, similar to a tubular steel pile. It may be driven into place and used as it is or it may be filled with concrete after driving. The West 'shell pile' is a proprietary design which uses shorter shell segments that are stacked onto a driving mandrel. Because of the relatively short length of the segments, the West pile is very useful in restricted headroom situations.

Whichever type and design of driven pile is used, there are essentially five possible ways to actually drive the pile, each of which adds its own variables to the process, some of which can have a direct influence on the potential for damage to the pile or adjacent structures during installation. The methods are as follows:

- Steam or air-driven drop-hammer
- Diesel-fueled internal combustion hammer
- Hydraulic driving hammer
- Hydraulically driven vibrator
- Direct-push installation

Design of an appropriate inspection and testing program must consider the selected driving method.

2.1.1 DROP-HAMMERS

The original drop-hammers were manually raised to the required height by rope on a either a winch or block-and-tackle and held by some sort of latch until released to drop onto the pile head. The introduction of steam power in the 1850s allowed several mechanical systems to be developed, ranging from a crude 'cat-head' similar to those used on modern geotechnical drilling rigs to perform the standard penetration test (SPT), to quite sophisticated systems with enclosed hammer pistons and a complex valve gear that enabled to operator to raise and drop the hammer piston with a high degree of precision, hence controlling both timing and drop height. The advent of efficient high-volume air compressors in the 1960s enabled operators to discard their steam boilers and convert their steam-powered hammers into more fuel and labor-efficient air-powered systems (Figure 2.2). According to David Redhead of BSP International Foundations, conversion of the BSP steam hammers to air power required only that the piston rings be replaced with a set that had a smaller gap in them

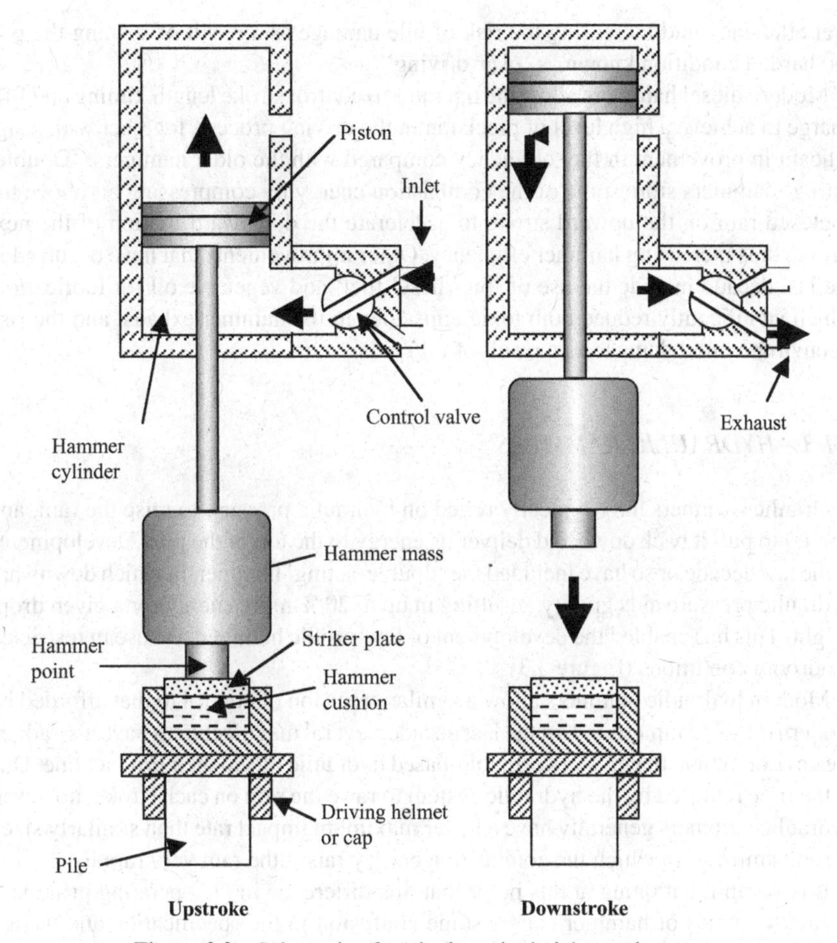

Figure 2.2 Schematic of a 'single-acting' air/steam hammer

(Hertlein, 2004a). Many air-powered driving hammers are in use today in various parts of the world.

2.1.2 DIESEL HAMMERS

The older models of internal-combustion diesel hammers were essentially just a heavy steel piston and cylinder with a simple fuel-injector, known as single-acting or 'open' hammers. As the hammer piston or 'ram' drops, it compresses the fuel/air mixture in the cylinder until it ignites. The combustion of the fuel forces the ram upwards and the reaction force pushes the cylinder down, driving the pile into the ground. The relatively crude machinery made precise control almost impossible, resulting in low fuel efficiency and increasing the risk of pile damage as a result of striking the pile too hard, a condition known as 'over-driving'.

Modern diesel hammers allow the operator to control stroke length, timing and fuel charge to achieve a high level of precision in the driving process, together with a significant improvement in fuel efficiency compared with the older hammers. 'Double-acting' hammers store some of the combustion energy by compressing air above the enclosed ram on the upward stroke to accelerate the downward motion of the next stroke, thus increasing hammer efficiency. Other improvements that have occurred in the last decade include the use of 'bio-diesel fuel' and vegetable oil for lubrication, which significantly reduce both toxic emissions in the hammer exhaust and the risk of environmental damage as a result of oil spills.

2.1.3 HYDRAULIC HAMMERS

Hydraulic hammers have typically relied on hydraulic pressure to raise the ram, and gravity to pull it back down and deliver its energy to the top of the pile. Developments in the last decade or so have included the 'double-acting' hammer, in which downward hydraulic pressure aids gravity, resulting in up to 20 % more energy for a given drop-height. This has enabled the development of low-profile hammers for use in restricted headroom conditions (Figure 2.3).

Modern hydraulic hammers allow a similar precision of control to that afforded by modern diesel hammers and, in the last decade, several manufacturers have also added the environmental benefits of vegetable-based hydraulic oils to their product line. Due to the time required by the hydraulic system to raise the ram on each stroke, however, hydraulic hammers generally have a lower maximum impact rate than similarly sized diesel hammers, in which the combustion energy raises the ram very rapidly.

It is worth mentioning at this point that this difference in the operating principles of the two types of hammer causes some confusion in the specification and testing procedures. It is a common misconception that diesel hammers are less efficient than hydraulic hammers, because the percentage of ram energy transferred to the pile in each drop is smaller. This apparent discrepancy is a result of the cushioning effect of the compressed air/fuel charge in the diesel hammer just prior to ignition. From a fuel consumption point of view, the diesel hammer ingests a small quantity of fuel just prior to the ignition point on each stroke, whereas the power-pack that provides the hydraulic hammer with its energy is typically driven by a multi-cylinder diesel engine that runs continuously between strokes. From a performance point of view, the diesel hammer is capable of delivering more strokes than a similarly sized hydraulic hammer in a given amount of time. The question of actual efficiency is therefore a complex one – by what parameter is efficiency judged? An understanding of the complexities of hammer behavior are important when analyzing apparent performance and evaluating test data. There is more discussion of this in Chapter 6 – High-Strain Testing.

Figure 2.3 Low-profile hydraulic hammer driving under low headroom. Reproduced by permission of American Piledriving Equipment, Washington, USA

2.1.4 PILE-DRIVING VIBRATORS

A pile-driving vibrator (Figures 2.4(a) and 2.4(b)) is typically attached to the top of the pile or casing by a set of hydraulic clamps. A set of hydraulically driven contra-rotating gears with eccentric masses are phased in such a way that the forces generated by the rotating eccentric masses are directed vertically up and down to vibrate the pile and the surrounding soil (Figure 2.4(c)). The contra-rotating masses are synchronized so that the centripetal forces that they generate oppose each other and so cancel out in the horizontal plane but combine in the vertical plane, creating a sinusoidal vertical vibration (Figure 2.4(d)). The vibrations reduce the soil friction on the sides of the pile, hence allowing the mass of the vibrator to push the pile down into the ground.

A disadvantage of older vibrators was that when the rotors were starting up (run-up) or slowing to a stop (coast-down), their speed would typically pass through a period of resonance with the soil or adjacent structures, where the frequency of the vibrations matches the natural or 'resonant' frequency of the soil or an adjacent structure. At

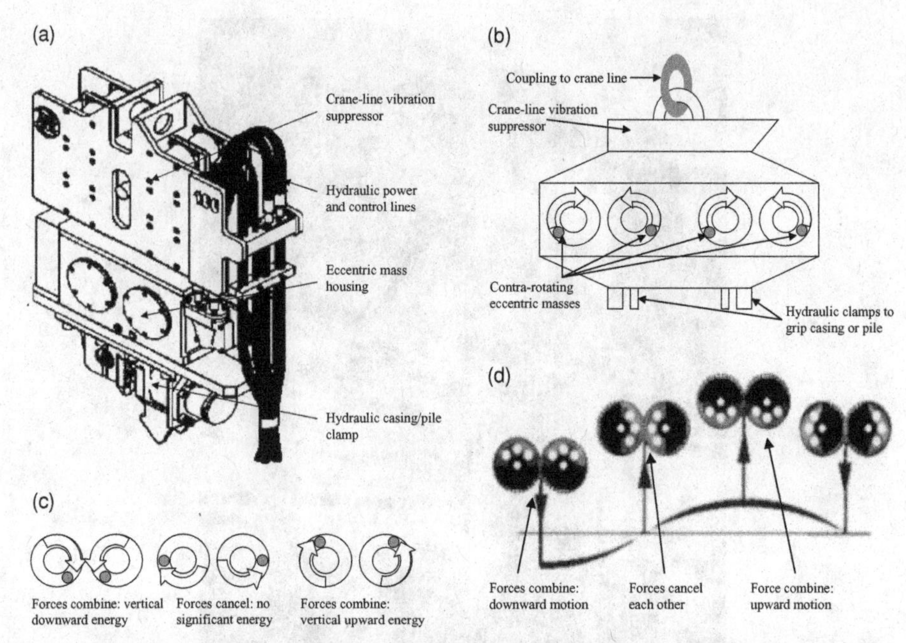

Figure 2.4 (a) Illustration and (b) schematic of a typical modern casing/pile-driving vibrator. (c) Vibration forces from one pair of rotating eccentric masses. (d) The effect of phase on rotating eccentric masses. Parts (a) and (d) reproduced by permission of American Piledriving Equipment, Washington, USA

resonance, the response of the soil to the energy input from the vibrator results in amplification of the vibrations. At resonance, vibrations in the soil can become highly perceptible, and even uncomfortable, to observers nearby. When structural resonance occurs, vibrations in the structure are also amplified and may easily reach levels that can cause damage to the structure. The reader may be more familiar with another commonly mentioned aspect of this phenomenon – where soldiers marching over a bridge are instructed to deliberately 'break-step' to avoid the risk of their marching causing resonance and resulting in uncomfortable or damaging vibrations in the bridge deck. A very graphic instance of this was demonstrated in London, UK, in June of 2000, during the opening of the Millennium Bridge over the River Thames. The cadence of pedestrians passing over the bridge matched the lateral resonant frequency of the bridge, with the result that the bridge started swaying from side to side by up to 70 mm (Arup, 2000).

Recent developments in pile-driving vibrator design include 'variable moment' systems, which allow control of the rotor eccentricity, so that no significant vibrations are developed until the machine is up to operating speed, thus avoiding most of the problems associated with soil or structural resonance during run-up or coast-down.

Figure 2.5 Stillworker press-in piling system used inside a building – note the steel weights stacked around the machine to increase reaction mass. Reproduced by permission of Thatcher Engineering, Illinois, USA

2.1.5 DIRECT-PUSH PILE INSTALLERS

A development that gained ground in the late 1990s is the direct-push or 'press-in' pile driving method for sheet piles, tubular micropiles and H-piles (Figure 2.5). The direct-push method relies on the mass of the machine to provide resistance for pushing as much of the first pile into the ground as possible with hydraulic rams. On some direct-push systems, the machine is then clamped to the first pile driven, adding that pile's uplift reaction to the mass of the machine, so allowing the system to push the second pile deeper. Depending on the configuration and required depth of penetration, the direct-push device may be clamped to several piles to provide the needed resistance for installation of long piles. Alternatively, subframes can be attached to the machine which can then be loaded with concrete or steel blocks to provide additional reaction mass. The low vibration levels and low headroom requirement for the direct-push equipment are distinct advantages on some urban sites.

2.1.6 ADVANTAGES AND LIMITATIONS OF DRIVEN PILES

Driven piles are often the most economical choice for certain types of foundations, largely because of the speed with which they can be installed and because little or no excavation of soil is needed, reducing both the amount of heavy equipment needed on site and the need to haul spoil away. This latter fact may be of particular importance on a site where contaminated soils are encountered.

A second advantage is that driven piles are relatively easy to install on an incline, or 'batter', to resist lateral forces or a combination of vertical and lateral forces. Purists in the current generation of foundation constructors and engineers frown on the term 'battered piles' in this context, but the name is a well-established and widely used colloquial term, unlikely to disappear in the near future.

Possible causes of problems when driving piles are the noise and the vibration. In many urban or residential environments, the noise levels caused by driving hammers are unacceptable to the local inhabitants. If the site is close to adjacent structures, the vibrations caused by either driving hammers or vibrators may cause damage to older buildings and problems with the operation of sensitive equipment such as large computer disk drives, microscopes and medical imaging equipment.

One of the most commonly quoted problems with driven piles is that there is really no way to be certain that they have been driven straight and true, nor to determine what they are founded in, other than comparing the length driven with the geotechnical exploration logs. There is at least one documented case where the tip of a steel H-pile emerged from the ground a few metres away from where it was being driven in! (See chapter 4 – Figures 4.1 and 4.2).

In addition, it is not possible to drive a pile into bedrock to form an embedded socket. Such rock sockets are important where there is a risk of floodwaters scouring out the soil around bridge foundations or where significant lateral forces must be resisted, such as in earthquake-prone areas, or for slope stabilization or landslide prevention.

2.2 CAISSONS AND DRILLED SHAFTS

Shafts that are constructed by excavating soil and rock and then filling the resultant hole with concrete are known by different names in different parts of the world. In Europe, the Middle East and Asia they are usually referred to simply as bored piles. In the Midwestern United States they are often called caissons. In the Southern and Eastern United States they are called drilled shafts, while in California they are called cast-in-drilled-hole (CIDH) piles. There are several different construction techniques and the choice of construction method will be governed by factors such as soil conditions, site access, number and size of shafts, budget and local availability or mobilization costs. The personal preference of the engineer of record is often also a deciding factor – a known method is often selected over a method with which the engineer has no personal experience, even if the cost is a little higher.

Hand digging of deep shafts is still practiced in some parts of the world, where low groundwater tables and cohesive soils make the practice possible. For most of the world, however, shaft excavation is done by machine. Smaller diameter shafts – up to about 1 m or so in diameter – may either be drilled by a rotary drilling rig equipped with an auger, or excavated by repeatedly dropping a special steel casing vertically to cut into the soil and then removing it to extract the plug of soil retained inside. The latter is known as the 'shell and auger' system. It is relatively slow, but the equipment is inexpensive and requires less headroom than a conventional crane or truck-mounted rotary rig. The small size of the shell and auger rig can be an advantage when working in an existing building to upgrade or underpin foundations.

Rotary rigs can drill shafts ranging from a few hundred millimeters up to more than four meters in diameter. Truck or carrier-mounted rigs fill the lower end of the diameter range, having the advantage of being relatively easy to mobilize and set up on site, but the dimensions and available power of these rigs limit the shaft size that can be constructed. Typically truck or carrier-mounted rigs are used for shafts up to about 30 m deep and up to 2 m in diameter.

Larger shafts usually require the use of crane-mounted rigs. The biggest of these are equipped with twin engines to power the drilling turntable and can excavate shafts in excess of 4 m diameter to depths of more than 70 m. The crane-mounted rig is more costly to mobilize and more complex to set up than a truck-mounted rig.

When drilling in cohesive soils with no water-bearing layers, the excavation may be unsupported throughout the construction process. Where there is risk of soil cave-in due to low cohesion, water-bearing soils or other unstable conditions, a steel casing may be inserted as the hole is advanced. Alternatively, using water or a slurry to create a head of hydraulic pressure that stops the inflow of groundwater can stabilize the excavation. The slurry may be a simple mix of water and native soil generated by the drilling process, but is more usually an engineered mix of water and bentonite clay or water and a polymer 'thickening agent'. Bentonite is a type of clay derived from volcanic ash that is dried and milled into a fine powder form, sometimes reformed into small pellets. It readily forms a thick slurry when mixed properly with water and allowed to hydrate. Bentonite slurry is a thixotropic material that will set to the consistency of thick yogurt when undisturbed, yet will readily return to fluid form when agitated. Bentonite slurry will hold small particles in suspension; thus, it can be circulated via holding and settling tanks to transport fine drill cuttings to the surface for disposal. Saline or brackish water causes bentonite to flocculate or 'clump', thus reducing its ability to hold back any inflow of groundwater or hold small particles in suspension. Therefore, in marine environments, attapulgite, another type of clay with similar properties to bentonite, but with less proclivity toward flocculation in salt water, is used instead. Attapulgite is often referred to in the drilling industry as 'salt gel'.

A recently developed alternative to the crane-mount is the pile-top rig, which uses the particle-suspension properties of slurry to transport cuttings out of the excavation. The pile-top rig is attached directly to the top of the casing, and uses a sectional drill-stem, where additional sections are added at the top as the hole is advanced downward. Unlike the truck or crane-mount rigs, which typically extract the cutting tool or auger every few feet to remove the cuttings or spoil, the pile-top rig uses reverse circulation, where an air-lift system continuously draws the slurry up the hollow drill-stem to flush the cuttings back to the surface, where the slurry is pumped off to settling tanks so that the cuttings can settle out and the slurry can be recirculated. The air-lift works by injecting compressed air into the hollow drill-stem just above the cutting head. As the air rises up inside the drill-stem, it reduces the overall density of the slurry column inside the pipe. The density differential forces the heavier slurry around the outside of the drill-stem up through holes in the cutting head, drawing the cuttings with it. The pile-top rig is becoming popular on deep-water marine projects, where keeping a barge stable enough to safely operate a crane-mount can be problematic.

Two other relatively recent developments in large-diameter drilled shaft construction technology are the casing oscillator and the casing rotator. As the names suggest, the oscillator twists a special thick-walled casing into the ground with a back-and-forth partial rotation, whereas the rotator turns the casing continuously. Both methods exert downward pressure, or 'crowd', on the casing for installation and upward pressure for removal. The bottom edge of the casing is fitted with cutting teeth capable of grinding through boulders and cutting into bedrock. The spoil is excavated with a clamshell bucket as the casing is advanced. Additional sections of casing are bolted on as necessary as the excavation advances. In rock, a drop-hammer or chisel is used to break up the rock core so that the clamshell bucket can extract it.

2.2.1 *ADVANTAGES AND LIMITATIONS OF DRILLED SHAFTS*

From the designer's point of view, one of the main advantages of drilled shafts is that the shaft can be designed to cope with a variety of load conditions, including downward axial, uplift, lateral and bending caused by seismic events in practically any combination. In addition, drilled shafts can be designed in a wide range of sizes ranging from a few hundred millimeters to more than three meters in diameter. A single drilled shaft can therefore replace multiple driven piles. Moreover, unlike driven piles, a drilled shaft can be extended into bedrock, thus enabling it to resist seismic forces and scour conditions better than driven piles.

During construction, one of the main advantages of drilled shafts over driven piles is that the excavated soil allows the contractor and the inspector to confirm soil conditions as the shaft is advanced, and so verify that the soil strata in which the shaft will be founded are as anticipated in the design.

Conversely, because drilled shaft construction requires the removal and disposal of the excavated soil, it can be problematic if the soils are contaminated with hazardous materials.

As with driven piles, the use of vibrators to install and remove temporary casing close to adjacent structures may cause problems with older structures and operation of sensitive equipment.

2.2.2 *ADVANTAGES AND LIMITATIONS OF SLURRY*

If slurry is required for the construction of deep foundations, the contractor must consider the properties of the various slurries when determining which type to use. Not only do the different types of slurry have different performance characteristics, but each requires different equipment and thus has an impact on the space required on site, the project schedule and the overall cost:

- Slurry construction usually requires storage or holding tanks for the slurry and pumps to circulate the material between excavation and the holding tanks as required.
- The mineral slurries, bentonite and attapulgite, must be mixed with water and allowed to hydrate for up to 24 h in order to reach the appropriate consistency, whereas the polymer slurries are almost 'instant' and the polymer can be added to the water in the excavation.
- The mineral slurries have a higher density than the polymer slurries, which may be advantageous under certain circumstances where high groundwater pressures must be balanced.
- Mineral slurries suspend fine particles much longer than polymer slurries. This necessitates the use of circulating pumps and desanding equipment to control the density of the mineral slurry.

Another point to consider when using mineral slurries is that suspended materials may settle out if there is a delay in placing concrete, creating a layer of debris at the base of the shaft that will show up in the results of integrity tests.

The reverse is true of polymer slurries. A typical polymer slurry, based on a polyacrylamide polymer, is anionic, meaning that it carries a negative electrical charge. Most soil particles carry a positive charge, which causes an attraction between the soil particles and the polymer chains of the slurry. The subsequent clumping or flocculation of the soil particles leads to rapid settlement of these particles to the bottom of the excavation, where they can be removed easily by techniques such as a clean-out bucket or air-lifting. Not only does this property eliminate the need for slurry desanding equipment, but the drilling contractor can take advantage of this by cleaning the base of the shaft after completion of drilling, and so reduce the risk of excessive debris piling up at the base in the event of delays in starting the concrete placement process.

When slurry is used to support the excavation, the question of disposal must also be considered. It is becoming increasingly difficult and expensive to dispose of mineral slurries such as bentonite or attapulgite in the United States, with many authorities classifying them as hazardous waste. Typical polymer slurries, on the other hand, can be effectively neutralized by adding household bleach or hydrogen peroxide. After a brief period to allow settlement of any suspended fine materials, what remains is basically just dirty water, and many agencies will permit disposal by either spreading it on the ground to evaporate or discharging into the sanitary sewer system.

2.3 DIAPHRAGM WALLS, CUT-OFF WALLS AND BARRETTES

The need for greater lateral stiffness under certain circumstance led to the design of the Barrette. Typically rectangular or oval in plan cross-section, the early versions were constructed by drilling two or more shafts that intersected to form a vertical slot in the soil. Modern equipment such as the 'Hydro-Fraise' from France uses a combination of hydraulically powered cutting wheels attached to a rectangular casing that incorporates either an air-lift or suction spoil removal system. An alternative system uses a clamshell bucket that opens to the required rectangular cross-section and can thus excavate the barrette in a series of 'bites'.

Diaphragm or slurry-trench walls are constructed with similar equipment, making multiple adjacent passes to excavate each panel. If the purpose of the wall is simply to form an impermeable barrier or cutoff wall, the design usually just calls for a series of contiguous rectangular panels. Where the wall is intended to be a retaining wall or support vertical or lateral loads when partially excavated, typical designs may incorporate a combination of rectangular panels and 'T-shaped' sections, known as counter-forts (Figure 2.6).

Cutoff walls in stable soils may also be constructed by drilling a series of interlocking, or 'secant' shafts. Typically, a row of primary shafts is drilled first, using medium-to-low-strength concrete (lean mix) spaced at about half of a diameter apart. Then, the secondary shafts are drilled in between the primary shafts. The secondary shafts cut into the primary shafts on either side, creating an interlocking barrier that can block fluid flow through the soil. If it is necessary to resist lateral soil pressure in

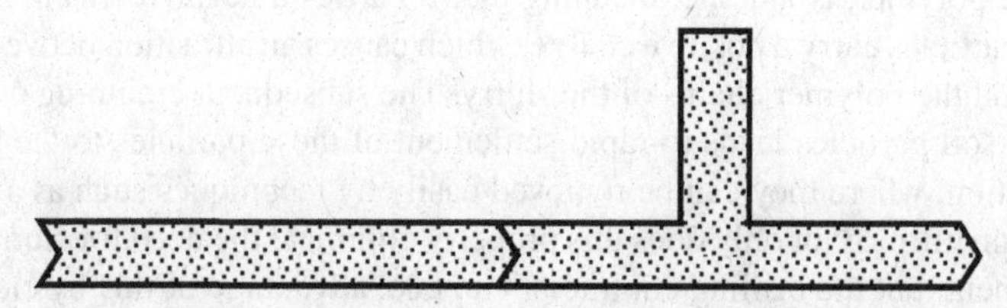

Figure 2.6 Cross-section of a diaphragm wall with a straight panel and counter-fort section

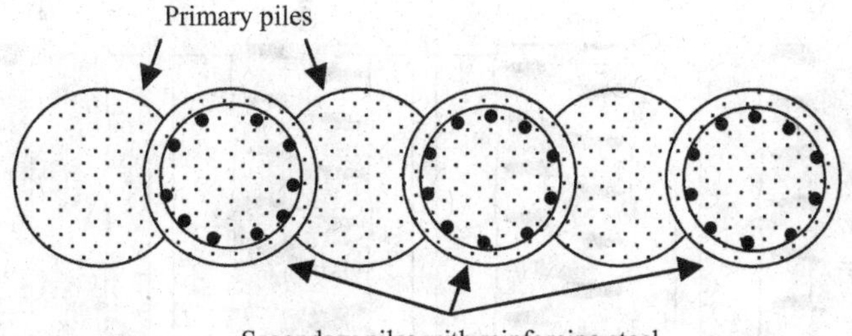

Figure 2.7 Cross-section of a secant pile wall

a retaining wall, or support load in a cut-and-cover tunnel or bridge, reinforcing steel can be installed in the secondary shafts (Figure 2.7). A steel H-pile has sometimes been used as an alternative to a reinforcing cage.

2.4 AUGERED, CAST-IN-PLACE PILES

Augered, Cast-in-Place (ACIP) is the American name for the piles known in Europe as Continuous Flight Auger (CFA) piles. The pile is constructed by screwing a hollow-stem auger into the ground. The exterior of the auger has a continuous spiral blade along its length-hence the European name. In the American practice, once the auger has reached the required depth, grout is pumped through the hollow stem as the auger is slowly withdrawn, still rotating. The spoil is thus removed and the soil is replaced with a column of grout (Figure 2.8). In Europe it is more common to use a 'micro-concrete', made with a mix of fine and coarse sands, or very small crushed aggregate. Micro-concrete is designed to be easily placed by small-diameter pump hose, yet achieve a similar strength to normal concrete.

Reinforcing steel can be placed in the shaft by pushing the reinforcing material down into the wet concrete or grout, or by placing a single high-tensile bar in the auger stem before drilling and pumping the grout or concrete around it. Placement of the reinforcing steel is also sometimes used as a quality control check. Where a single bar is to be placed in the center of the shaft, an oval frame or 'basket' of lighter-gauge steel is formed around the bottom end of the reinforcing bar. The purpose of the basket is to ensure that the reinforcing steel is placed in the center of the pile, but it can also be sized such that it can be used to check that there has been no soil squeeze or 'necking' of the hole (Figure 2.9).

A recent variant of the standard ACIP shaft is the 'displacement ACIP' shaft, in which most of the soil is not removed. Instead, the tip of the auger is designed to force the soil outward, compacting it, and the grout is placed under higher pressure than for the standard ACIP pile.

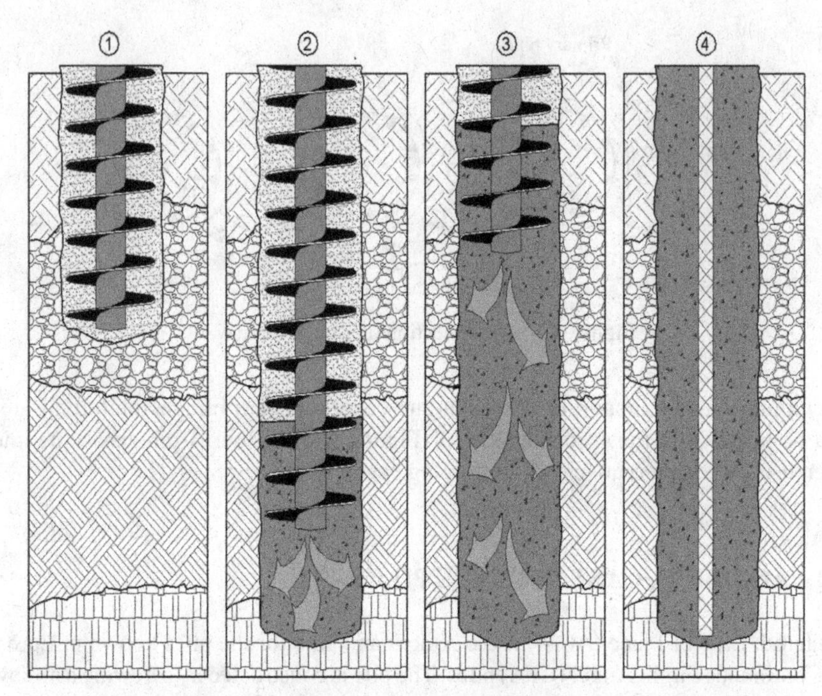

Figure 2.8 The four principal stages of constructing an augured, cast-in-place pile. Reproduced by permission of West Coast Foundation, Florida, USA

2.4.1 ADVANTAGES AND LIMITATIONS OF ACIP PILES

Under certain circumstances, ACIP piles can offer many of the advantages of drilled shafts, plus lower noise and vibration levels. The displacement ACIP pile also significantly reduces the amount of spoil that needs to be handled and/or disposed of. On the other hand, the soil displaced by a large group of displacement piles can result in significant heave or uplift of the ground surface, which can create unintended tensile forces in the shafts.

Inspection of ACIP piles is much more difficult than for drilled shafts, and the construction process is highly operator-dependent. Until recently, the inspector had to rely on counting grout pump strokes, monitoring pump pressure and looking for grout return on the auger flights. Recent developments in drill rig instrumentation, coupled to computer-based monitoring systems, have greatly improved the inspection process by providing continuous data on such critical measurements as grout placement rate, tip pressure, auger rotation and auger withdrawal rate. The output from these systems is typically a graphical representation of the approximate shape of the constructed shaft. The technology has proven sound, but at the time of writing this manual, reliability is still an issue for some of the transducers that must be used. When the monitoring system fails, the engineer must fall back on more traditional inspection methods, or consider using nondestructive integrity tests after the grout or concrete has set.

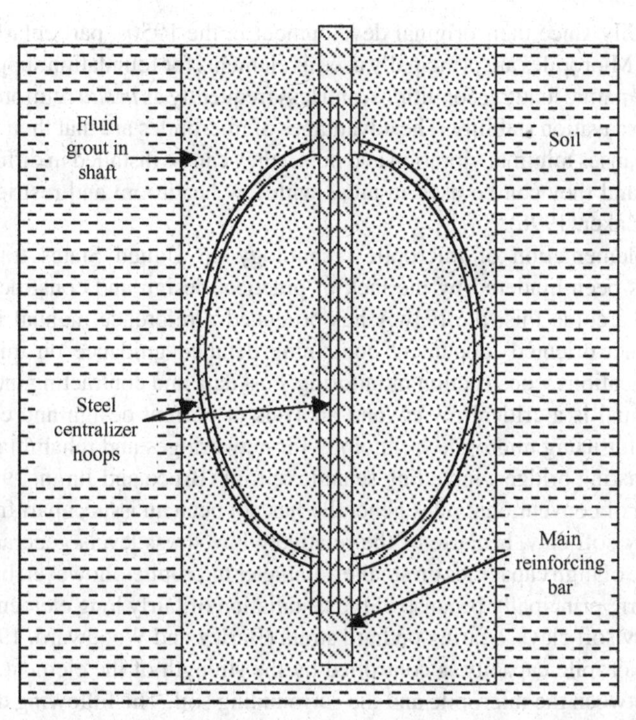

Figure 2.9 'Football' centralizer for insertion of a single reinforcing bar

Depending on the quality and variability of the local soils, ACIP piles can have very irregular cross-sections. This can make the test data from the surface impact methods, such as Impulse Echo or Impulse Response, difficult to interpret and often inconclusive. Installation of the access tubes for the downhole methods is also problematic, particularly if there is no full-length reinforcing cage. Some success has been achieved by placing a single access tube in the auger stem before grouting or concreting the shaft.

2.5 MICROPILES OR MINIPILES

It is worth spending a little time here to get a better understanding of the micropile. Partly because it has several names around the world – micropile, pin-pile, minipile, reticulated pile, root pile – it is a foundation construction method that is poorly understood by the civil engineering community in general. As of 2004, the industry leaders have agreed that 'micropile' is the preferred term for these foundations. Most members of the deep foundations NDT community are also relatively unfamiliar with micropiles, for reasons which will be discussed shortly.

A micropile is a small-diameter pile (typically less than 300 mm), either a jacked steel tube or a drilled and grouted reinforced pile. Worldwide use of micropiles has grown steadily since their original development in the 1950s, particularly since the mid-1980s. Micropiles are principally used as elements for foundation support to resist static and dynamic loading conditions, and less frequently as *in situ* reinforcements for slope and excavation stability. The advantages of micropiles are that their installation procedure causes minimal vibration and noise, they can be installed in difficult ground conditions and they can be used in areas with low headroom and restrictive access (Hayword Baker, 1996).

The implementation of micropile technology on United States transportation projects has been hindered by the lack of practical design and construction guidelines. To overcome this, the FHwA has produced a reference manual intended as a 'practitioner-oriented' document containing detailed guidance on micropile design, specifications, construction monitoring, cost data and contracting methods. The FHwA manual is a critical review and analysis of current design and construction techniques for using micropile technology for new bridges and rehabilitation and/or repair of existing bridges, including seismic retrofit (Bruce and Juran, 1997).

Buckling of conventional piles generally has not been an issue, apart from slender piles in very soft clay, because of the relatively low stress levels generally applied. With the use of high capacities on smaller-diameter micropiles, however, buckling can be of concern, principally because of the installation of fairly long elements (>15 m) through very soft clays, bearing on hard strata (Barley and Woodward, 1992). In this case, the micropile capacity is controlled by the strength of the element rather than the bond between the micropile and the surrounding soil. The following descriptions and classifications of micropiles were first published in the FHWA report by Bruce and Juran, and are reproduced by kind permission of Dr. Bruce. See also Bruce, 1994.

2.5.1 *APPLICATIONS*

These can be classified as follows:

(1) For structural support
   - Foundations for new structures
   - Seismic retrofitting
   - Underpinning of existing foundations (Figure 2.10)
   - Repair/replacement of existing foundations
   - Arrest/prevention of movement
   - Upgrading of foundation capacity.

(2) For in situ reinforcement
   - Embankment, slope and landslide stabilization
   - Soil strengthening and protection
   - Settlement reduction.

The effect of groups and networks is discussed by Benslimane et al. (1997). The interested reader is also referred to Bruce and Nicholson, 1989.

Figure 2.10 Jacking-in a steel tube micropile for structural underpinning. Reproduced by permission of Thatcher Engineering, Illinois, USA

2.5.2 *DRILLED MICROPILE TYPE/CLASSIFICATION*

2.5.2.1 Classification Based on Behavior

Case 1

The pile resists the applied loads directly. This is usually true for cases where individual piles or groups of piles are used. In this context, a group is defined as a tight collection of piles, each of which is subjected to direct loading. When axially loaded piles of this type are designed to transfer their load only within a remote founding stratum, pile head movements will occur during loading, in proportion to the length and composition of the pile shaft between structure and founding stratum.

Case 2

Reticulated root pile structure. This concept is to support and stabilize using an interlocking, three-dimensional network of reticulated piles, and involves the creation of a laterally confined soil/pile composite structure that can work for underpinning, stabilization and earth retention (Juran et al., 1996).

2.5.2.2 Classification Based on Method of Grouting

Type A

Grout is placed in the pile under gravity head only. Since the grout column is not pressurized, sand–cement mortars as well as neat cement grout may be used. The pile drill hole may have an under-reamed (belled) base to aid pile performance in tension (Taylor, et al., 1996).

Type B

Neat cement grout is injected into the drilled hole as the temporary steel drill casing or auger is withdrawn. Pressures are typically in the range of 0.3 to 1 MPa and are limited by the ability of the soil to maintain a grout-tight seal around the casing during its withdrawal, plus the need to avoid hydro-fracturing pressures and/or excessive grout consumption.

Type C

Neat cement grout is placed in the hole as for Type A. Between 15 and 25 min later, before hardening of this primary grout, similar grout is injected, once, via a pre-placed sleeved grout pipe at a pressure of at least 1 MPa.

Type D

Neat cement grout is placed in the hole as for Type A. Some hours later, when this primary grout has hardened, similar grout is injected via a pre-placed, sleeved grout pipe. In this case, however, a packer is used inside the sleeved pipe so that specific horizons can be treated several times, if necessary, at pressures of 2 to 8 MPa.

In addition to these classes, the jacked or pushed pipe pile (Fig 2.10) must be considered a sub-type of micropile, since the consensus of opinion is to include pin-pile and minipile under the generic term “micropile”.

2.5.3 RELATIONSHIP BETWEEN MICROPILE APPLICATION, DESIGN CONCEPT AND CONSTRUCTION TYPE

These aspects are summarized in Table 2.1.

Table 2.1 Relationship between micropile application, design concept and construction type.

Application	Structural support	*In situ* earth reinforcement 1	2	3	4
Sub-application	• Underpinning existing foundations	Slope stabilization and excavation support	Soil strengthening	Settlement reduction	Structural stability
	• New foundations	—	—	—	—
	• Seismic retrofit	—	—	—	—
Design concept	Case 1	Cases 1 and 2	Case 2 with minor Case 1	Case 2	Case 2
Construction type	• Type A (bond zones in rock or stiff clay)	Type A (Cases 1 and 2) and Type B (Case 1) in soil	Types A and B in soil	Type A in soil	Type A in soil
	• Types B and D in soil (Type C only in France)	—	—	—	—
Estimated percentage of total applications	Probably 95 % of total worldwide applications	0–5 %	Less than 1 %	Not known	Less than 1 %

2.5.4 DESIGN ASPECTS

These can be listed as follows:

(1) Analytical viewpoint – settlement, bursting, buckling, cracking and interface considerations:
(2) Practical viewpoint – corrosion protection and compatibility with the existing ground and structure:
(3) Economical viability.

2.5.4.1 Mode of Load Transfer

The system must be capable of sustaining the anticipated loading requirements within acceptable settlement limits. Drilled and grouted micropiles typically transfer load to the ground through skin friction, as opposed to end bearing for steel tube micropiles: a pile 200 mm in diameter with a 5-m-long bond zone has a peripheral area 100 times greater than the cross-sectional area. This mode of load transfer directly impacts performance in that the pile movements needed to mobilize lateral friction resistance are of the order of 20 to 40 times less then those needed to mobilize end bearing.

2.5.5 NONDESTRUCTIVE TESTING

Integrity testing of micropiles by nondestructive methods has not proven to be very successful to date, for the following reasons:

- The relatively small pile diameter and construction methods nearly always preclude the installation of tubes for down-hole testing.
- The very great pile length/diameter ratio usually prevents sensible results from stress-wave tests applied at the shaft head, such as Sonic Echo and Sonic Mobility.

2.5.6 RESEARCH AND DEVELOPMENT

A five-year National Project termed 'FOREVER' (Fondations Renforcées Verticalement) has been undertaken by a French consortium under the aegis of the Institute for Applied Research and Experimentation in Civil Engineering (IREX) in cooperation with the USA Federal Highway Administration (Bruce *et al*., 1997). The project conducted under the technical direction of Professor Francois Schlosser and Dr Roger Frank of the French National Civil Engineering School (ENPC) involves research institutes, contractors and governmental agencies.

FOREVER includes desk studies, numerical modeling, laboratory testing (centrifuge) and full scale field-testing. Its chief objective is to promote the use of micropiles in all fields: deep foundations of new buildings and structures, stabilization of slopes and embankments, underpinning of existing foundations, and seismic retrofitting of retaining walls and shallow foundations.

2.6 STONE COLUMNS AND OTHER SOIL IMPROVEMENT TECHNIQUES

There are a number of techniques for providing structural support that are really soil improvement techniques rather than foundation construction methods *per se*, but they are worth mentioning here because they are often included as part of a complete foundation construction package and several of them have been significantly improved in the last decade as a result of site experience and equipment development. Some of them can also have a significant effect on the performance in nondestructive tests.

The four techniques most widely used are stone columns, injection grouting, deep soil mixing and dynamic compaction. As with the basic foundation construction methods, there are several proprietary variations of each method. Only the basic principles, and their implications for nondestructive testing, will be discussed here.

2.6.1 STONE COLUMNS

Stone columns are constructed by driving a casing into the ground, either as a displacement method or with subsequent excavation of the soil within the casing. The stone or crushed rock is then placed within the casing in a series of placements or 'lifts' and consolidated by tamping or vibration as the casing is gradually withdrawn with each lift.

At the time of writing this book, there is no effective NDT method for assessment of stone columns, because the unbonded nature of the stone creates a myriad of interfaces that scatter and disperse any stress waves generated on or in the shaft. Borehole radar, however, has evolved significantly in the last few years, creating the possibility that a directional antenna could be effective in profiling a stone column from an adjacent borehole where the soil conditions are amenable to radar usage. Other possible techniques are discussed in the following sections.

2.6.2 DEEP MIXING

Deep mixing may be performed to create discrete columns, linear wall-like structures or grid-like masses. In each case, the principle is the same. A set of mixing blades mounted on a hollow stem are screwed into the ground to the required depth. The binding agent – lime, cement, flyash or a mixture of them – is pumped through the hollow stem to be mixed with the soil as the mixing blades rotate and are gradually withdrawn.

The effectiveness of deep soil mixing is highly dependent on the nature of the soils involved and the available moisture. One of the present authors recently worked on a project in the United States where deep soil mixing was used to stabilize weak soils and help control groundwater for a drilled shaft project. The soil mixing was generally successful, but in some locations the resulting material was so hard that the drilling contractor could not auger through it as planned, and had to resort to core drilling.

To date, NDT methods for assessment of soils improved in this manner have met with mixed success, largely due to the often variable nature of the finished product. There is, however, a glimmer of hope offered by current research based on tomography of cross-borehole radar and electrical resistivity data. For further information, please refer to the section on 'Cross-borehole Radar and Resistivity Tomography' in Chapter 13 of this book – Current Research.

2.6.3 *PERMEATION GROUTING*

Permeation grouting, as its name implies, consists of pushing a nozzle into the soil and forcing grout through it to increase the density of weak soils or to fill voids and displace water from them.

An important variation of the grout injection method that is widely used in Europe, and occasionally used in the United States since the 1970's, is known by its original French name – Tube á Manchette. This grouting method is typically used to improve the soil around the base of a drilled shaft. Access tubes for the sonic logging test (see Chapter 10) are linked across the base of the shaft to form several sets of U-tubes. The cross-links at the base of the shaft have a number of holes drilled in them and are clad with a neoprene sleeve. The latter prevents the ingress of concrete and retains the water during the initial construction and testing phases of the project. Once the shaft has been proven sound and accepted, grout is pumped down the access tubes, to force off the neoprene seal, and consolidate the soil around the base of the shaft.

Significant increases in bearing capacity have been achieved with this method, but the effective increase in shaft cross-section caused by the grout can make interpretation of NDT test results very difficult. Use of surface tests, such as the Impulse-Response or the Impulse-echo methods (see Chapters 7 and 8), is likely to be inconclusive on shafts that have been consolidated with the Tube á Manchette method.

2.6.4 *DYNAMIC COMPACTION*

Essentially, dynamic compaction consists of a crane dropping a large mass onto the ground to consolidate it by squeezing out the air and moisture. The dynamic compaction rig moves over the site in a grid pattern, dropping the mass as many times as is necessary to achieve the desired consolidation. While the technique is a 'pure' soil improvement method, the vibrations caused by dropping the mass have the potential to cause problems with adjacent foundations through a combination of unanticipated soil consolidation at adjacent sites, lateral soil pressures and cracking of the immature concrete of newly constructed deep foundations nearby.

If dynamic compaction is specified for a construction site close to a concurrent deep foundation construction program, vibration monitoring and integrity testing of the new foundations are highly recommended by the authors of this present book.

3

How Soils Affect the Choice of Foundation Type

Deep foundations exist because of the need to overcome difficult soil conditions. The earliest timber piles were placed in marshy ground or floodplains to give house dwellers dry and safe lodgings, above the influence of fluctuating water levels and bog denizens. Up until the 20th Century, the soil profile immediately beneath structures limited their size. A combination of mankind's technological ingenuity and growth of construction power has seen the development of many different deep foundation methods to overcome problematic soils. These technological advances continue to be driven by the need to develop sites with unfavorable soil conditions as the best sites are being used up.

In spite of these developments, many deep foundation problems are still with us today, principally as a result of site soil conditions or local environmental limitations. Examples such as the selection of the wrong pile type for the soil profile, unforeseen soil and groundwater conditions during construction, and errors in selection of soil parameters for the foundation design are still common. Damage to neighboring structures caused by vibration or excessive soil settlement is also a distressingly frequent occurrence.

Local knowledge of the 'right' foundation for the soil conditions in populated regions is typically considered in local building codes, while the experience of local deep foundation designers and contractors is built into these codes. However, increasing federalization of engineering practice throughout the world (starting with large construction groups in such places as the USA and Europe) has sometimes negated this experience. It is now common for deep foundation contractors, offering their large arsenal of foundation construction methods, to bid for and win projects far outside their normal zone of experience. In these instances, the contractor relies completely

Nondestructive Testing of Deep Foundations B.H. Hertlein and A.G. Davis

upon the thoroughness of the geotechnical site investigation which precedes the bid to identify soil and groundwater issues that control foundation type selection and identifies complications that may arise during construction. Supplementary claims for expenses incurred by unexpected or changed conditions are often made after the contract has started, citing incorrect or missing information in the site investigation. Sometimes, however, it is simply that the contractor is working outside his normal sphere of experience, lacks local knowledge and therefore misses or underestimates the significance of certain items or features noted in the geotechnical reports. Typical errors include under-estimating the strength of local rock formations or the quantity of large boulders in the soil. Either condition will certainly result in extra drilling time and may even force the contractor to change the drilling equipment.

One striking example of how soil conditions affect the development of deep foundation construction techniques is the contrast between deep foundation practices in The Netherlands and in France in the building and rebuilding boom that followed World War II. The low-lying, alluvial terrain in The Netherlands has resulted in urban and industrialized areas being concentrated around waterways and ports, with considerable depths of soft and loose compressible alluvial soils, and water tables at the ground surface. Driven piles were traditional, starting with timber and then proceeding to steel and later to pre-cast concrete as design loads became heavier. With these increasing loads, one of the most economical and sure foundation types in The Netherlands was the pre-cast concrete driven pile in its various forms. France, on the other hand, saw development of most urban zones in areas where soils ranged from soft to very stiff, with frequent changes in both vertical and horizontal soil profiles. Very stiff soft rock layers, as well as cobbles and boulders, are common, and as a result, driven piles encountered difficulties. The increase in design loads and the need to build on unfavorable sites led French engineers to rely more on bored piles (drilled shafts), with rapid development of techniques that utilized drilling mud or slurry to keep holes open during drilling and concrete placement in cohesionless or water-bearing soil strata.

There are many other examples of regional development in piling or deep foundation drilling techniques to suit local soil and environmental conditions. One often used by geotechnical instructors is that of belled (under-reamed, or enlarged base) drilled shafts, which were relatively common in the early days of hand-dug shafts, with laborers digging to the founding stratum and then expanding the shaft base diameter to exploit the harder soil or rock conditions at that level. Over time, foundation engineers in urban regions identified hard strata at fairly constant depths that could be used as high-capacity bearing layers, and local codes were expanded to allow for bells at shaft bases. Particularly good examples of this now commonly practiced technique are to be found in Chicago, USA and in London, UK. The accepted practice in Chicago fifty years ago was to dig straight-sided drilled shafts down to bear on the hard limestone at between 30 and 40 m below grade. It was eventually realized that a stiff boulder clay layer (known locally as 'hardpan') was present over most of the Chicago city center area at a depth of 20 to 25 m below grade, and that using belled shafts founded on this layer could generate considerable savings in construction costs.

Similarly, with the appearance of high-rise construction in the London area, particularly in the Docklands redevelopment projects of the 1970s and 1980s, foundation design engineers had to increase the working loads of the traditional bored piles in the London Clay. The stronger strata, such as the Woolwich and Reading Beds (dense sandy gravel), below the London Clay then became the level at which expanded base shafts could be founded.

The modern techniques for excavating the bell at the base of a shaft are fully mechanized and do not resemble the hand-dug methods first used to build this type of foundation. However, it is still relatively common to see inspectors going down to the base of open shafts to inspect the cleanliness and shape of the bell, although engineers are gradually being dissuaded from this practice by national safety regulations. In a dry hole, inspection by a 'down-hole camera' is a viable alternative to exposing a person to the potential dangers of cave-ins and pockets of deadly gases.

Driving piles is a relatively quick and inexpensive way of constructing deep foundations but, quite apart from the problems encountered when trying to drive through boulder clays, pile driving also typically causes considerable ground vibration and airborne noise. These effects are often unacceptable in urban environments and on sites close to vibration-sensitive industrial or medical facilities. In those situations, drilled shafts or augered-cast-in-place (ACIP) piles are preferable, even though they may be more costly.

Similarly, installing and removing temporary casing for drilled shafts often requires the use of a vibratory driver. If vibration is a significant concern, such as close to a historic structure or a vibration-sensitive process, and the soils would require stabilization by temporary casing for a drilled shaft, then ACIP piles may be the only viable solution for relatively small-diameter shafts.

If large-diameter shafts are required in such a situation, the Casing-Oscillator or the Casing-Rotator techniques may be suitable. In these techniques, a temporary heavy-duty casing, fitted with cutting teeth around its bottom edge, is installed in the ground by either fully or partially rotating (oscillating) it while providing a downward force, or 'crowd'. The spoil is extracted from inside the casing by a clamshell excavator. On completion of concrete placement, the casing is withdrawn hydraulically by reversing the crowd of the machine, and continuing to rotate or oscillate it. These methods produce relatively little ground vibration or airborne noise, compared with pile driving or casing vibration.

Soils containing large boulders can be problematic for both driven piles and drilled shafts but can be drilled relatively easily by the Casing-Rotator or Casing-Oscillator methods. Boulders or sections of rock too large for the clamshell to extract are broken up by drop-hammer or percussion drill.

Permanent casings may be required to stabilize granular soils with relatively high water flow or to extend foundation shafts up through the water for bridge piers, jetties and similar structures.

Last, but by no means least, are the design requirements. Will the structure be required to resist severe lateral loading from high winds, marine wave action or vessel impact? Will the structure be subject to seismic events that may liquefy surrounding soils or impose lateral loads? If the potential for scour during flooding exists, then substantial depths of the surrounding soils may disappear altogether, while the flood waters and floating debris simultaneously push against the foundations. Similarly, if the upper soil layers are too soft or loose to provide the necessary resistance against relatively small-diameter driven or ACIP piles when the expected lateral loads are applied, then large-diameter drilled shafts or rectangular barrettes may be the only feasible solution.

Thus, by the time all the soil, design requirements and environmental issues are considered, and the relative costs of the various methods are compared, the foundation designer often finds that the choice of foundation type has been dictated by site circumstances, required foundation performance and the project budget. What remains for the engineer to do is to calculate the necessary layout, quantity and dimensions.

As will be seen later in this book, the selected foundation type and construction procedure also limit the engineer's choices when it comes to inspection and testing methods, including nondestructive testing.

4

Traditional, Visual and New Inspection Methods for Deep Foundation Construction

Traditional deep foundation quality control has been primarily visual. The degree of inspection or material testing that is possible has always been dependent on the foundation type and was often minimal. The following paragraphs are not intended to be a guide to deep foundation inspection, but rather to point out some of the inherent limitations of the traditional methods. Later sections show how nondestructive test techniques complement the visual observations of the inspector by compensating for some of the limitations of visual inspection.

4.1 DRIVEN PILES

Driven piles can be visually inspected prior to installation, allowing the inspector to verify pile length, the condition of any joints used and the overall integrity of the shaft. In the case of steel or timber piles, this is usually adequate. In the case of concrete piles, however, mishandling of the pile by the crane or forklift operator when unloading the delivery truck or picking up the pile to start driving can cause multiple fine cracks that are practically invisible to the naked eye, but can rapidly worsen under the impact of the driving hammer.

An experienced inspector who keeps a close eye on the work at all times is likely to notice when the pile is being mishandled, but all too often the inspector is not present when the piles are first delivered to the site and unloaded.

Nondestructive Testing of Deep Foundations B.H. Hertlein and A.G. Davis

Figure 4.1 'A strange object emerges from the ground!'. Reproduced by permission of Bengt Fellenius Consultants, Alberta, Canada

During pile driving, the inspector can verify pile plumbness, or verticality, monitor the rate of advance and determine if the pile has been 'set' by counting the number of hammer blows for a given increment in depth, or verify the final depth achieved by measuring the residual length of pile sticking up out of the ground.

In the event that a pile is advancing more slowly than expected, it is difficult for the inspector using only visual means to determine if the slow advance is due to stiff ground conditions or to an inefficient hammer. Likewise, damage can occur in the pile if the hammer strikes it too hard, or 'over-drives' it, but the unaided inspector is unlikely to spot such an occurrence unless the damage causes a sudden significant difference in the movement of the pile. In addition, the assumption is made that the pile is being driven straight. In reality, particularly with steel piles, the tip of the pile can wander off-line and there is no way that the inspector can determine where the tip finally ends up. A rather extreme but highly illustrative example was reported by Dr Bengt Fellenius, who showed that one particular driven H-pile made a complete 'U-turn' and emerged from the ground surface several meters behind the pile-driving rig (Figures 4.1 and 4.2).

4.2 AUGERED, CAST-IN-PLACE PILES

The inspector of augered, cast-in-place (ACIP) piles can only check the length, diameter and straightness of the auger prior to drilling and monitor the plumbness of the

Figure 4.2 'You can never be sure where the end of a steel pile is going!'. Reproduced by permission of Bengt Fellenius Consultants, Alberta, Canada

drilling rig as the hole is drilled. Once the auger has reached the required depth, the inspector can monitor the flow of concrete through the pump by counting pump strokes and watching the pump pressure gauge, and observe the cuttings that are brought to the surface on the auger flights. The inspector should also be watching auger rotation speed and rate of withdrawal – both critical functions during grout placement.

When the shaft is close to completion, the inspector can watch for the first return of grout to the surface on the auger flights, since adequate immersion of the auger tip in the fluid grout is also critical to maintaining adequate grout pressure and thus creating a sound shaft.

With so many things to monitor, and the noise of the drilling rig, the grout pump and the ready mix truck to contend with, it is quite possible for a momentary fluctuation in grout pressure to go unnoticed by either the operator or the inspector. The result of that fluctuation, however, could be a significant reduction in shaft cross-section, or 'neck-in'.

It has been generally accepted in the industry that the quality of ACIP shafts is very dependent on the skill and experience of the driller. The inspector is typically in the role of observer and has little or no influence on the work itself. Largely because of this, the last ten years or so have seen the development of several automated measuring systems designed to aid the operator by giving real-time feedback of critical system parameters and to aid the inspector by providing documentation of parameters that could only be estimated before. While the systems are proprietary, with some minor differences, they essentially perform the same functions. Grout pressure and flow rate are monitored continuously, as is auger rotation and withdrawal rates. Because the grout volume that has been placed and the position of the auger tip are both known accurately, the computer controlling the system has enough information to calculate and draw the probable shape of the shaft that has been created.

4.3 DRILLED SHAFTS

A drilled shaft inspector has traditionally been able to observe and measure much more than the driven pile or auger-cast shaft inspector. Beginning with the contractor's tools, the drilled shaft inspector can verify that the temporary and permanent casings are of the correct diameter, are undamaged and have not become oval in cross-section. Soil augers, cleaning buckets and belling tools can be checked for appropriate dimensions and correct operation.

As drilling progresses, the inspector can monitor the spoil being removed from the excavation and compare it with what is expected from the geotechnical report for the site. In this way, any unexpected changes in soil conditions that may be detrimental to the shaft construction process or performance under load can be recognized immediately and appropriate steps can be taken to accommodate the changed conditions. The inspector can also verify that the bearing stratum upon which the foundation design is based has been reached.

4.3.1 DRY HOLE CONSTRUCTION

The inspector can verify the plumbness and the depth of the shaft using a weighted tape measure. If no casing has been used and the hole is dry, the sides of the bore can be inspected using a powerful flashlight. If the hole is dry and cased, the inspector can enter the excavation to inspect the condition of the base and the sides of any rock socket, although this practice is now frowned upon in many parts of the world. The dangers of toxic or suffocating gases and caving soils have been amply demonstrated in the recent past and the general consensus of the deep foundations industry is that down-hole entry should be avoided whenever possible. Experience has shown that even when a hole is drilled in the dry and purged with clean air, soil can unexpectedly cave from the sides of a bell or around the bottom of a temporary casing, to allow a sudden inrush of methane or some other noxious or suffocating gas.

One of the major reasons for down-hole entry in the Midwest United States is the often-seen specification requirement for a pilot hole at the base of rock sockets in karst or heavily fissured limestone formations. The pilot hole is usually only a few inches in diameter. The inspector is expected to enter the main shaft and use a rod with an angled spike at the tip to scrape the sides of the pilot hole to detect any significant discontinuities or voids (Figure 4.3). The last decade has seen the

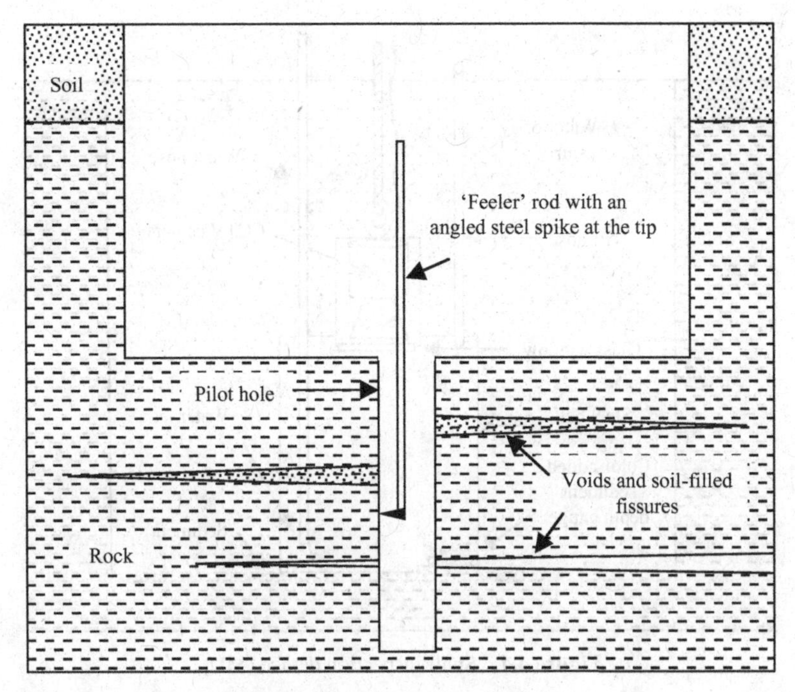

Figure 4.3 'Feeling' a pilot hole

development of inexpensive explosion-proof high-resolution video cameras that can be lowered down the shaft and so eliminate the need for personnel entry in most cases. In the last few years, miniaturized versions of these cameras have become available; so making remote visual inspection of small-diameter pilot holes a practical alternative to down-hole entry.

When the reinforcing cage is installed in a dry hole, the inspector can verify that the cage does not scrape the sides and cause soil cave-in. When the cage is set, the inspector can verify visually that it is concentric within the shaft and not buckled or bent.

During concrete placement in a dry hole, the inspector can watch to ensure that the falling concrete does not damage the reinforcing cage and in an uncased hole the inspector can verify that no soil has caved into the hole during concrete placement.

4.3.2 WET HOLE CONSTRUCTION

If the hole is wet, however, the inspector's options are severely restricted. The presence of water or slurry in the hole makes it impossible to check the condition of the sidewalls or the cleanliness of the bottom by direct observation. Similarly, installation of the

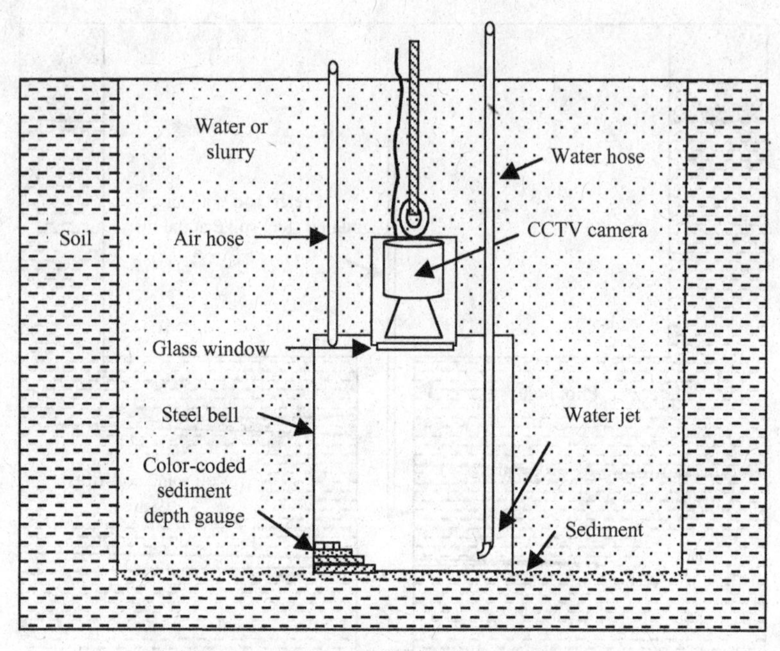

Figure 4.4 Shaft inspection device (SID)

reinforcing cage in a wet hole can be observed only at the top of the shaft. Cage straightness and concentricity cannot be verified visually, neither can soil cave-in caused by scraping the sides.

A recent addition to the inspector's 'toolkit' in the United States has been the Shaft Inspection Device, or 'SID' (Figure 4.4). This device and a smaller version called the 'Mini-SID' (Figure 4.5) were developed in Florida by the Schmertman and Crapps geotechnical company, specifically to permit a visual assessment of the cleanliness of the bottom of any shaft drilled under water or slurry. SID is essentially a miniature version of the old-fashioned diving bell, equipped with a closed-circuit television camera, air and water supplies, and a color-coded sediment gauge (see Figure 4.4). The air is used to blow water or slurry out of the bell when it has been lowered to the bottom of the shaft. The water jets can be used to dislodge or stir up sediment and rock cuttings caught under the bell, while the sediment gauge allows rapid assessment of the depth of any sediment when the excavation has been allowed to stabilize.

A tool that promises to be helpful in inspection of wet holes is the ultrasonic borehole caliper, also known variously as an ultrasonic borehole profiler or drilling monitor. At the time of writing this book, there are only two proprietary systems that are commercially available for large-diameter drilled shafts. One is an analog system that is manufactured in Japan by Koden, while the other is a digital system that is manufactured in the United States and marketed by R&R Visuals.

Figure 4.5 'Mini-SID' ready for deployment down a shaft. Reproduced by permission of Applied Foundation Testing, Inc., Florida, USA

The analog system, sold as a drilling monitor, has been used primarily in the Asian construction market. There has been some limited specification of the ultrasonic borehole caliper for projects in the United States in the last two or three years and the California Department of Transportation (CALTRANS) purchased an analog system in 2002. The general feedback from users of ultrasonic borehole calipers in the USA has been that it is a slow procedure, and also rather expensive. A significant disadvantage to operators who are used to digital storage of data and post-acquisition processing for analysis is that the current Koden drilling monitor system is analog, with data recorded directly onto thermal print paper. There is no data storage with the system, and so the operator must take great care to ensure that all data are accurately recorded on the thermal print-out in the field.

The digital system recently developed in the United States does store the data, and in post-processing there are a number of image-format options available to the user, including three-dimensional wire frame models and color-rendered three-dimensional images. At the 2004 DFI NDT seminar at the University of Cincinnati, R&R Visuals exhibited a computer program that even enables the user to 'fly' down the interior of the virtual shaft (Figure 4.6).

According to a senior engineer at CALTRANS, the primary benefit of the drilling monitor is the ability to determine the profile of test shafts before performing the load test. This information can enhance the analysis of load-test data, particularly when determining load distribution, and so increase confidence in the test results

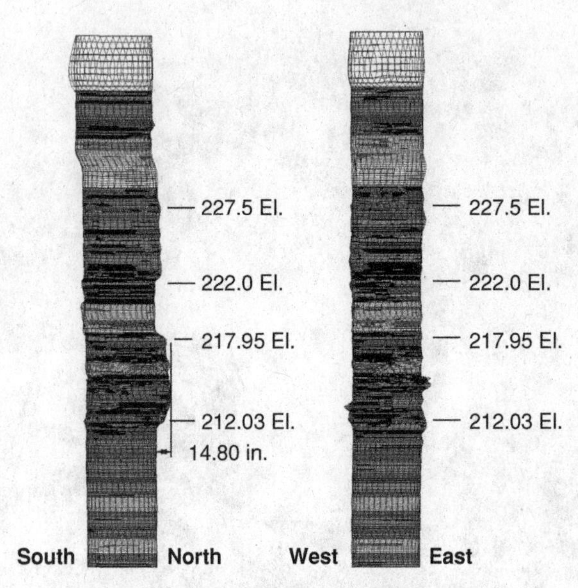

Figure 4.6 Wire-frame image generated by an ultrasonic borehole calliper. Reproduced by permission of R&R Visuals, Inc., Indiana, USA

(Hertlein, 2004b). The current ultrasonic caliper systems are generally regarded as being too time-consuming to be economically practical on production shafts unless cross-section dimensions and/or rock-socket integrity information are vital to the design engineer, such as when working in karst terrain. At least one other company manufactures a similar tool for assessment of small-diameter core holes in concrete and rock, however, so if demand for the tool rises, it is likely that other manufacturers will enter the market, the technology will advance and prices will drop.

In a wet hole, concrete must be placed from the bottom up using a tremie pipe. The inspector can only estimate or count the quantity of concrete being placed and measure the concrete level in the hole with a weighted tape measure. Keeping a graph of volume placed against depth reached can help the inspector identify unusual conditions such as significant over-size (overbreak) and significant cave-in of soil.

Measuring or estimating the volume of concrete placed, however, can be problematic. If the concrete is being placed by pump, and the volume of the pump cylinder is known, then simply counting the number of pump strokes can give a reasonably accurate estimate of total volume placed provided that the pump doesn't miss any strokes. Similarly, when relatively small (1 yd^3 or so) buckets or skips are used to place the concrete, it is a simple matter to count the number of buckets. However, when concrete is placed using larger skips, or is dumped direct from the chute of the ready-mix truck, estimates of volume placed are usually much less accurate and are very heavily influenced by the experience of the inspector.

Once removal of any temporary casing begins, the inspector can sometimes get into a position where the reinforcing cage can be watched for movement, but more usually it can only be checked once the casing is completely withdrawn. If the cage has moved, it is an indication of a possible anomaly, and some form of nondestructive test of the concrete is usually warranted when it has hardened.

4.4 THE INSPECTOR'S ROLE

For the purposes of this book, a discussion of the foundation construction inspector's role is a logical link between traditional inspection techniques and testing, both for full-scale capacity determination and for nondestructive integrity verification. There are two reasons for this:

(1) There is a pervasive misconception that the inspector can be replaced by nondestructive testing.
(2) The importance of the inspector's observations and notes is typically underestimated by those unfamiliar with nondestructive testing.

To address these points in turn, the first fact that must be emphasized is that nondestructive testing was never intended to, and definitely should not, be used as a substitute for an experienced inspector. Full-scale load testing of every shaft is a practical and economic impossibility, and no single nondestructive test provides enough information to be used as a stand-alone quality verification tool. Unfortunately, on projects with tight budgets or unexpected cost over-runs, it is tempting to reduce the amount of money spent on inspection if it is known that the foundations will be tested by one of the nondestructive techniques. In reality, an experienced inspector can often identify problems in the construction process, allowing the contractor to correct them early in the project schedule, whereas nondestructive testing is often performed after construction is complete. All too often, this approach results in identification of multiple anomalies in the completed foundation, most of which could have been avoided if an experienced inspector was employed on the project – a clear case of 'too little, too late'.

The second point is that much of what the inspector sees and notes in the daily logs may become critically important if nondestructive testing locates an anomaly in a foundation. The inspector's notes may include information about the soil strata observed during excavation, delays or other problems noted during the concrete placement procedure, or anomalies in the pile-driving process. This information will almost certainly be important when trying to determine the nature and most likely cause of anomalies in the test results and deciding what modifications should be made to the construction process to prevent their recurrence.

In order to be able to take accurate and useful notes, the inspector must have a good understanding of the foundation construction process in question. When assigning a foundation inspector, it is tempting for some companies to save money by sending out young 'engineers-in-training', fresh from college, or field technicians who are relatively new to the job, rationalizing the decision by saying 'It will be good experience for you' or 'Hey, you only have to watch what's going on, and make a few notes'. Unfortunately, the notes taken by such a novice will probably have little relevance to the actual construction process, and will be of little use in determining the nature of an anomaly or, in the worst case, supporting a claim in court or arbitration.

To distill the essence of a seminar regularly presented by the present authors for the ADSC and the DFI, the inspector must:

- Understand the intent of the foundation design – i.e. end-bearing or side friction.
- Know the Unified Soil Classification System (USCS) or a locally recognized equivalent.[1]
- Know the intended foundation depth or bearing stratum.
- Be familiar with the soil exploration data (geotechnical boring logs) for the site.
- Know the proposed construction process for the foundation in question.
- Be familiar with the equipment normally used for the proposed construction process.
- Be competent and concise in oral and written communication.
- Be present for the entire construction process.
- At the end of each working day, compare notes with the foundation contractor's superintendent and reconcile any differences in apparent observations.

One of the most important steps is that which should be taken at the end of the working day, when the inspector and the contractor should get together and compare notes. Any discrepancies between them can be resolved while the events of the day are still fresh in everyone's mind. This apparently simple step can prevent misunderstandings and disagreements later on, and so help avoid the development of an adversarial relationship between the inspector and the contractor – a situation that helps nobody and can seriously hinder the inspector in the efficient performance of his or her work.

This list is only a summary of the main capabilities and duties of the foundation construction inspector – more detail can be found in the various inspector's manuals produced by the ADSC and DFI for drilled shafts, augered, cast-in-place piles and driven piles (ADSC-IAFD/DFI International, 2003; DFI International, 2003). The inspector must meet these criteria if the observations and notes recorded by the inspector during the construction process are to have the maximum value when it comes to determining the nature of anomalies identified by nondestructive tests, plus estimating their likely significance.

It is also important to understand that driven piles, augered, cast-in-place piles and drilled shafts are each very distinct foundation types with their own specific potential problems. Just because an inspector or a testing engineer can boast of twenty years experience with driven piles, it does not necessarily mean that he or she has any

[1] There are several soil classification systems in use around the world, although the most widely used is the USCS. The inspector must be familiar with the one that applies to the work site in question.

knowledge of the drilled shaft construction process. Similarly, an experienced drilled shaft inspector may be completely unfamiliar with the unique challenges encountered during the construction of an ACIP pile, or the care required when handling a pre-cast concrete pile. When selecting either an inspector or a testing agency, qualifications should be carefully reviewed to ensure that they are relevant to the foundation type in question.

5

A Review of Full-scale Load-testing Techniques

Deep foundation static load testing began as a way of checking that driven timber piles could carry their working loads. Simply, a load equal to the working load, plus an equivalent safety-factor load, would be placed on the pile for a given time and the pile settlement would be measured. The load was made up of the heaviest material available close to the site, usually stacked on a wooden platform, and was referred to as 'kentledge'. Since working loads were relatively light, such a test was within the reasonable capabilities of constructors, with easily constructed loading systems. Specifications were developed for test-load increments and the duration of their application, often as a function of soil type. When other pile materials such as steel and reinforced concrete appeared, these specifications were extended to those pile types.

Static load tests on working piles are always disruptive to the smooth progression of a piling site and in the third decade of the 20th Century engineers turned to the use of dynamic pile formulae such as the *Engineering News Record* formula to check predicted pile capacity. These methods used the measured pile driving 'set' with an estimate of the energy transmitted to the pile from the hammer for each hammer blow. The resultant cost saving was significant, and even today it is rare to see static load testing on working driven piles. This approach has its drawbacks, for the immediate pile/soil dynamic response to driving is not necessarily that of pile behavior under a maintained, static load. This is particularly true for soft cohesive soils, where pile driving gives neither information on future pile settlement under load, nor any possible increase in shaft friction over time. Various procedures, such as pile re-driving after a certain delay, have been devised to compensate for this. At the present time, the development of the correlation of measured dynamic pile behavior during driving

Nondestructive Testing of Deep Foundations B.H. Hertlein and A.G. Davis

with static pile performance has narrowed the gap between dynamic driving and static performance prediction and methods such as CAPWAP that measure the energy transmitted to the pile and the soil response to the hammer blow using gauges attached to the pile head are in common usage today (see Chapter 6 – High-Strain Testing).

However, only some form of load test can check the design of bored piles in their various forms after pile construction. With the constant increase in pile size and pile design loads over the last forty years, this has become more difficult and expensive. Much of the effort in the development in full-scale load-testing methods over the last decade has addressed the need to test drilled shafts designed for loads often in excess of 2000 tonnes.

A static load test is made usually for one or other of the following reasons:

- To determine the pile load–settlement relationship, particularly in the region of the anticipated working load.
- To serve as a 'proof test' to ensure that failure does not occur before a load is reached which is a selected multiple of the chosen working load.
- To determine the actual ultimate bearing capacity as a check on the value calculated from dynamic or static formulae, or to obtain information that will enable other piles to be designed by empirical methods.

In the first two cases, testing up to a 'proof load' can be performed on working piles after construction, usually at an early stage in the contract. This proof load is usually twice the design working load. However, full benefit from load testing is obtained from pre-contract shafts constructed to obtain the maximum information possible, using carefully planned loading procedures well beyond working loads and extra instrumentation in the shafts to follow the development of load transfer to the soil during loading.

O'Neill and Reese, in their treatise on 'Drilled Shaft Construction Procedures and Design Methods', sponsored by the United States Department of Transportation and the Federal Highways Association (FHwA), make a comprehensive review of field loading-test procedures recommended for highway bridge drilled shaft foundations in the USA, and their treatise can be considered to represent the state-of-the-art for the drilled shaft industry in the world at the time it was written (O'Neill and Reese, 1999). Because of the differences in foundation types, construction procedures, local soil conditions and legal requirements, different codes and recommendations for static load testing are applied in different countries. For example, the United Kingdom refers to the British Standard BS 8004 (British Standards Institution, 1986) for recommended load-testing practice. These recommendations differ in certain ways from those given by O'Neill and Reese and similar differences can be found in codes of practice for pile load tests in other countries. However, there is now a movement to reduce these differences, with the development of European and International Codes. In this text, the approaches defined in O'Neill and Reese are referred to, with certain comments on practices in other countries where considered significant.

5.1 STATIC LOAD-TEST TECHNIQUES – AXIAL COMPRESSION

Before load testing a drilled shaft, it is essential to establish whether the test is for proof-loading the shaft or to take the pile to 'failure'. The latter is defined as either plunging of the shaft under small incremental loads, or a gross settlement or lateral deflection of 5 % of the shaft diameter if plunging has not been experienced. Tested shafts must reproduce production shaft shapes, sizes and construction procedures as closely as possible, and be constructed in the same soil profile. Shafts for proof testing are not normally instrumented, apart from the external load-cells and gauges required to establish load–displacement plots. On the other hand, pre-contract load tests can include instrumentation in the shaft to measure load distribution with depth, measuring load transfer through the shaft sides in order to confirm design assumptions about shaft/soil load transfer. Lateral load tests can also include instrumentation in the shaft to measure shaft-deflection profile with depth. This means that pre-contract tests can be of the order of ten times as expensive as contract shaft proof tests and are rarely undertaken. Loads during testing are usually applied in stages (increments), with the load at each stage maintained for a specified time or incremental deflection. The usual procedure is for each load increment or decrement to be held for the time required for the rate of shaft-head displacement to reduce to a specified level.

5.1.1 REACTION SYSTEMS

Static load tests require a reaction system constructed over the shaft for load application. Before the 1950s, kentledge was the only practical way to apply this load. Kentledge consists of building a loading frame or platform immediately above the shaft to be tested, and then placing dead weights on the frame to act as the reaction for the jacks applying the test load (Figure 5.1). Care has to be taken that the soil supporting the kentledge has enough capacity to support this load. This system is efficient when anticipated test loads are relatively low, as is the case for proof-loading of smaller-diameter shafts. However, the advantage of using larger-diameter shafts to carry heavier loads became apparent in the 1950s, together with the greater use of pre-contract load testing, and reaction loads increased dramatically. The need to provide in excess of 1000 tonnes reaction load became commonplace and kentledge was not always the solution.

To resolve this dilemma, reaction shafts were introduced to provide these higher loads. Reaction (or slave) shafts are constructed symmetrically about the test shaft and a reaction frame assembled above (Figure 5.2). Normally, either two or three reaction shafts are required for stability and their combined uplift resistances must be greater than the intended test load. Naturally, the zone of influence of the reaction shafts must be well away from the loading influence zone of the tested shaft. An added advantage of this approach is that the reaction shafts are usually the first to be built on

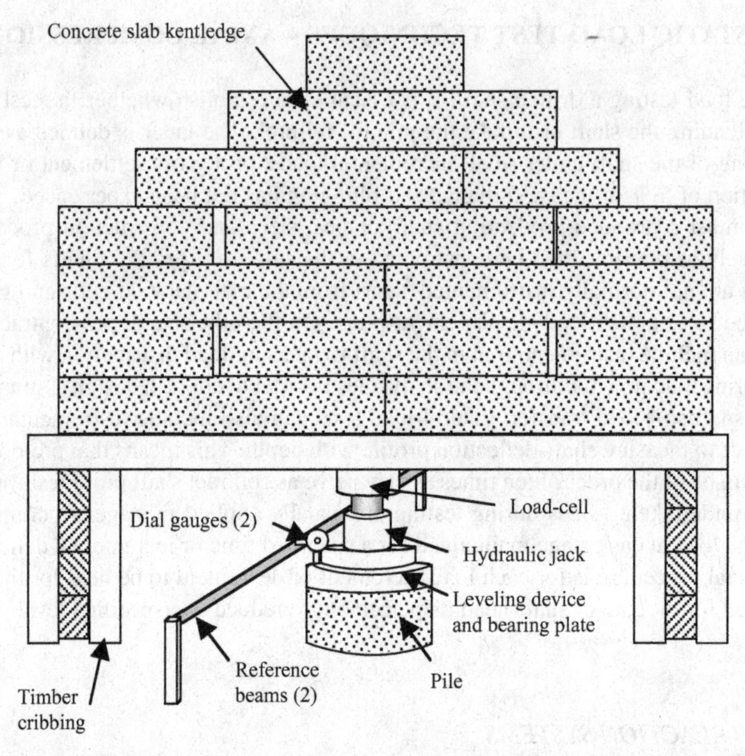

Figure 5.1 Schematic of a static axial compression load test with a 'kentledge' reaction mass (note that the 2nd reference beam has been omitted for clarity)

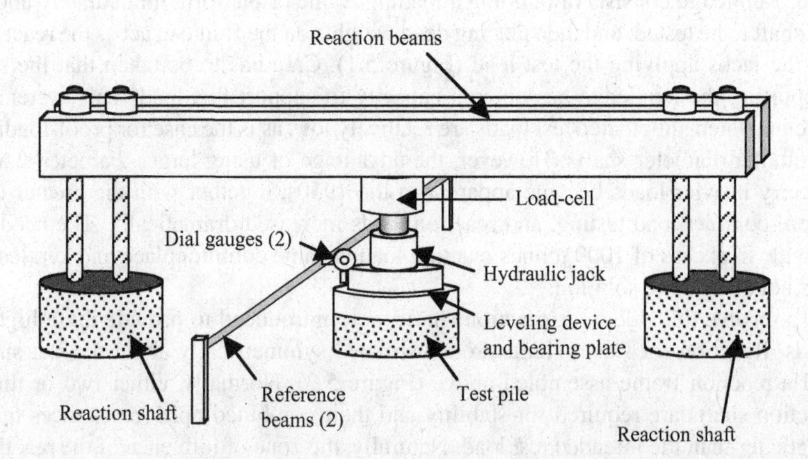

Figure 5.2 Schematic of a static axial compression load test with reaction shafts (note that the 2nd reference beam and gauges have been omitted for clarity)

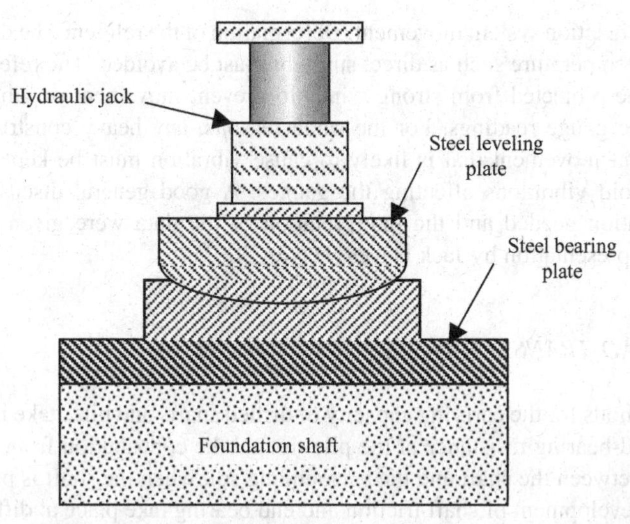

Figure 5.3 Typical load-test jack bearing/leveling plate combination

the site and afford the contractor the opportunity to test his construction techniques before test or contract shafts are placed. Any initial 'teething problems' can then be resolved before the contract begins. An alternative to reaction shafts can be provided in certain cases by using inclined high-strength ground anchors. Ground conditions, however, rarely make this method feasible.

In addition, at high loads it becomes very important to ensure that the force of the jack is uniformly distributed across the head of the shaft. The bearing plate typically incorporates a partial ball-joint feature that allows the jack to swivel slightly to compensate for if the top surface of the shaft and the underside of the kentledge or reaction frame are not perfectly parallel (Figure 5.3).

5.1.2 PROOF TESTING

Minimum required measurements during proof testing are the load and deflection at the shaft head. The simplest method of measuring load is to measure the hydraulic pressure in the loading ram by using a calibrated pressure gauge. The accuracy of this approach can be compromised by factors such as hydraulic loss, air in the system or by lack of ball joints or swivel-heads in the linkage between the ram and the shaft head to reduce the effect of ram friction. A better approach is to include a load-cell between the ram and the shaft head. Both load-cell and jack pressure can be read during the test.

Either dial gauges or linear variable differential transformers (LVDTs) are used to measure deflections of the shaft head. These gauges are attached to a reference beam independently mounted on supports away from any ground movements resulting from test shaft or reaction system movement. Deformation of the reference beam caused by changes in temperature such as direct sunlight must be avoided. The reference beam must also be protected from strong winds to prevent movement or vibration from affecting the gauge readings. For the same reasons, any heavy construction work or equipment movement that is likely to cause vibration must be kept far enough away to avoid vibrations affecting the gauges. A good general discussion of the instrumentation needed and the interpretation of the data were given in a recent conference presentation by Jack Hayes (Hayes, 2002).

5.1.3 LOAD-TRANSFER TESTS

Design methods for the carrying capacity of drilled shafts and piles take into account both the end-bearing resistance at the pile tip and the contribution from the friction generated between the shaft and the surrounding soil when the shaft is placed under load. The development of shaft friction and end bearing take place at different times during loading at the head of the shaft, and sometimes it is necessary to confirm design assumptions for these two factors by carrying out load-transfer-loading tests. These tests are considerably more expensive than the simpler proof tests described above, because of the need for designing and installing relatively sophisticated monitoring instrumentation in the pile shaft before construction is completed. Consequently, they are performed much more infrequently than proof tests.

The earliest method for measuring the development of load transfer with increasing loads on the shaft head was that of 'telltales'. These are typically unstrained or spring-loaded rods riding in vertical tubes set in the shaft before concreting. The telltale rods are placed in pairs of equal length; each pair extends to a different depth in the shaft, with the longest pair reaching nearly to the base of the shaft. When load is applied, differential displacement transducers operating between the shaft head and the telltale rods can measure the shaft shortening over a particular distance. Typical accuracies for these transducers are of the order of 0.0025 mm (0.00001 in).

As the shaft head load is increased, the compression along the pile shaft as a function of depth down the shaft can be plotted and the slope of this curve gives the unit strain in the shaft as a function of depth. In this way, the internal shaft load at any depth can be calculated by multiplying the axial shaft stiffness (shaft cross-sectional area × shaft concrete Young's modulus) by the strain observed at that particular depth.

The load distribution down the shaft can also be measured by a 'sister bar'. This is a section of reinforcing steel, usually between 1 and 1.2 m long, with a strain transducer mounted at its middle. The sister bar is tied to the shaft reinforcing cage and the lead wires routed to the surface. Vibrating wire transducers are usually preferred to electrical resistance gauges because of their greater stability over long time periods. However, electrical resistance gauges can be read more rapidly, with a large number of gauges read simultaneously using a small personal computer. Sister bars are the most prevalent instruments for reading shaft-load distribution.

For a full description of the placement and operation of telltales and sister bars, together with the advantages and the difficulties of their use, the reader is referred to O'Neill and Reese (1999) and/or Hayes (2002).

A typical loading procedure is that which is described in ASTM D1143, 'Standard Test Method for Piles Under Static Axial Compressive Load.' The load is applied in increments of 25 % of the design load. At each increment, the load is held steady for up to 2 h while foundation movement is noted. If settlement is not excessive, then the next increment is applied. Once the full test load is applied, it is held steady for at least 12 h, then removed in decrements of 25 % of the total test load. At each decrement, foundation movement is monitored for 1 h. A full static load-test cycle can thus take up to 32 h to complete. Since some specifications require at least one reload cycle, it can often take more than a week to complete a static load test, plus the time needed to construct and then remove the load-test equipment.

5.1.4 QUICK LOAD TEST

The duration of the standard static load test coupled to the disruption that it can cause to site activity can make it a very costly procedure. For certain types of foundation and soil conditions, an abbreviated load test may be appropriate. The 'Quick Load Test Method for Individual Piles' is given as an option in Section 5.6 of ASTM D1143. In essence, this option permits applying the test load in increments of 10 to 15 % of the design load, at intervals of 2.5 min. Once the full test load is achieved, it is held steady for 5 min, and then removed in a single step. The Quick Load Test can thus be completed in less than 1 h. It must be borne in mind, however, that, while the Quick Load Test provides excellent results in granular soils, the method cannot assess the long-term behavior of foundations in cohesive soils susceptible to creep. In certain cohesive soils, the Quick Load Test can significantly over-predict the capacity of the shaft, resulting in excessive settlement of the working shafts if used incautiously.

5.1.5 CONSTANT RATE OF PENETRATION TEST

The constant rate of penetration (CRP) test was developed in the United Kingdom after Whitaker at the Building Research Establishment Station established that CRP model pile tests in London Clay gave identical results to maintained load tests (Whitaker, 1963).

The CRP test progressively increases the compressive axial force to cause the pile to penetrate the soil at a constant rate until pile failure. Essentially, this test is applied to preliminary test piles or research projects only. Execution of the method is rapid and soil consolidation or creep is not allowed to take place. As a result, the load–settlement curve is easy to interpret. The British Standard BS 8004 2 states that penetration rates of 0.75 mm min^{-1} are suitable for friction piles in clay and 1.5 mm

min^{-1} for piles end-bearing in granular soil. The CRP test is not used for checking compliance with specification requirements for maximum settlement at given stages of loading. Estimating the maximum reaction loads required is also difficult, since the failure load is not known with any certainty before the test is undertaken.

5.1.6 BI-DIRECTIONAL LOAD TEST (OSTERBERG CELL)

The Osterberg cell, named after its inventor, Jorj O. Osterberg, entirely replaces the traditional jack and reaction frame system. The shaft load is applied through an expandable jack and load-cell cast within the test shaft, attached to the reinforcing cage (Figure 5.4).

The Osterberg cell is very simple mechanically, and consists of a metal piston and cylinder that create an expandable chamber holding pressurized fluid (oil or water). The piston and cylinder are each welded to a 50-mm-thick steel plate whose larger diameter approximates that of the test shaft. The pressurized fluid acts on the piston, and since the piston is usually at least 800 mm (32 in) in diameter, the Osterberg cell can apply relatively large loads for low hydraulic pressures. Loads of more than 2700 tonnes (3000 US tons) can be reached with the largest cells.

Once the concrete has hardened, the jack is pumped up, creating both upward and downward forces in the shaft. Depending on where the cell is placed in the shaft, load tests can be performed to separate and measure several important parameters. For example, if the Osterberg cell is placed at the bottom of a shaft, the upward force

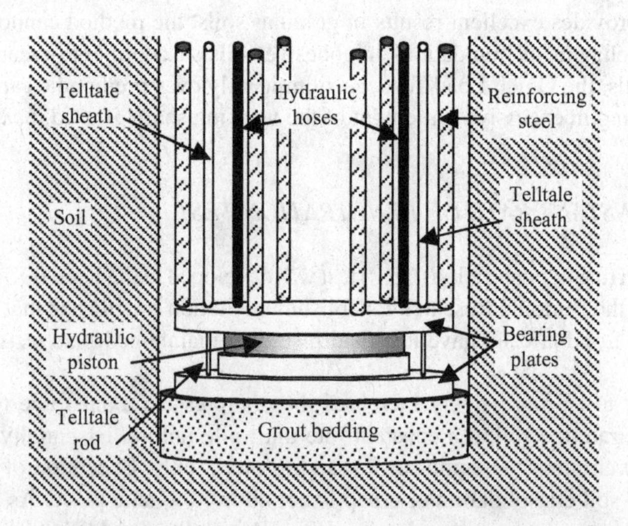

Figure 5.4 Schematic of an Osterberg cell load-test jack (note that the typical helical reinforcing steel has been omitted for clarity)

enables measurement of the side friction on the shaft, while the downward force enables measurement of the end-bearing capacity of the soil.

If the cell is placed at the bottom of a rock socket, the side friction in the socket can be evaluated. In soft soils where no appreciable end-bearing is expected, the Osterberg cell can be placed somewhere towards the mid-portion of the shaft, and thus use the side friction on the lower portion of the shaft for reaction against which to create the upward force in the upper portion of the shaft. The effective side friction can thus be measured in both portions.

For larger-diameter shafts, multiple Osterberg cells can be installed between the same pair of bearing plates (Figure 5.5). The largest capacity generated in this manner at the present time is more than 31 350 US tons (278 MN). News of this was released in the ADSC-IAFD Journal *Foundations Drilling* during the writing of this book (ADSC-IAFD, 2005). It would have been a monumental task, if not a practical impossibility, to construct a kentledge of a similar capacity.

The Osterberg cell has also been employed on its side to generate horizontal loads that were used to assess the elastic behavior of rock sockets in cases where the lateral stiffness of the rock is questionable or of critical importance to the shaft design (O'Neill and Person, 1998).

Another advantage of the bi-directional load test is that if the shaft tested is a production shaft and is proven acceptable, the Osterberg cell can be flushed out and pumped full of grout to become an integral part of the shaft, thus qualifying as a

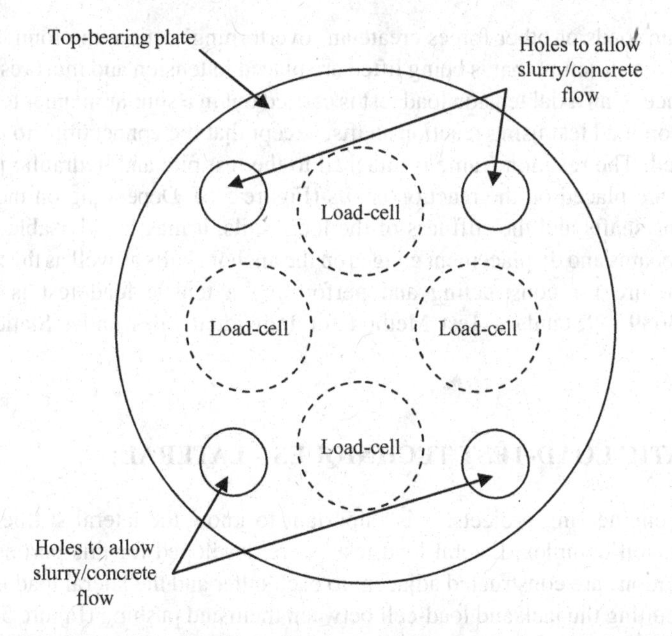

Figure 5.5 Plan-view of multiple Osterberg cells for large-diameter shaft load tests

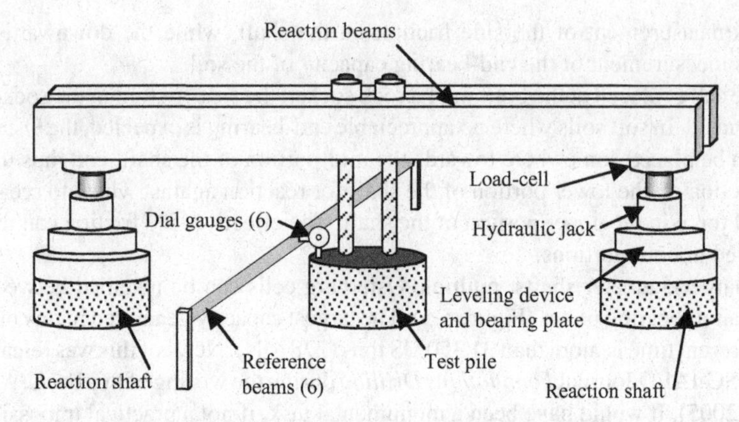

Figure 5.6 Schematic of a static axial tensile load test with reaction shafts (note that the reaction shaft reference beams have been omitted for clarity)

truly nondestructive test technique. For a more general discussion of the Osterberg cell method, the reader is referred to the 1992 FHWA report by the inventor, Dr Jorj Osterberg (Osterberg, 1992) or the paper by Hayes and Simmonds (2002).

5.2 STATIC LOAD-TEST TECHNIQUES – AXIAL TENSION

Where wind loads or other forces create an 'overturning' action, the foundations on the side of the structure that is being lifted are placed in tension and must resist uplift. Performance of an axial tension load test is carried out in a similar manner to the axial compression load test using reaction shafts, except that the connections to the shafts are reversed. The reaction frame is attached to the test pile and hydraulic jacks and load-cells are placed on the reaction shafts (Figure 5.6). Depending on the number of reactions shafts and the stiffness of the local soils, it may be advisable to install reference beams and displacement gauges on the anchor shafts as well as the test shaft.

A procedure for constructing and performing a tensile load test is given in ASTM D3689 – 'Standard Test Method for Individual Piles under Static Tensile Load'.

5.3 STATIC LOAD-TEST TECHNIQUES – LATERAL

For some engineering projects, it is important to know the lateral stiffness of the foundation/soil complex. Lateral load tests were developed for this purpose. Often two foundations are constructed adjacent to each other and the lateral load is applied by either putting the jack and load-cell between them and pushing (Figure 5.7(a)), or constructing a tension frame around the pair with the jack and load-cell on the outside of the reaction shaft, so pulling the two shafts together (Figure 5.7(b)).

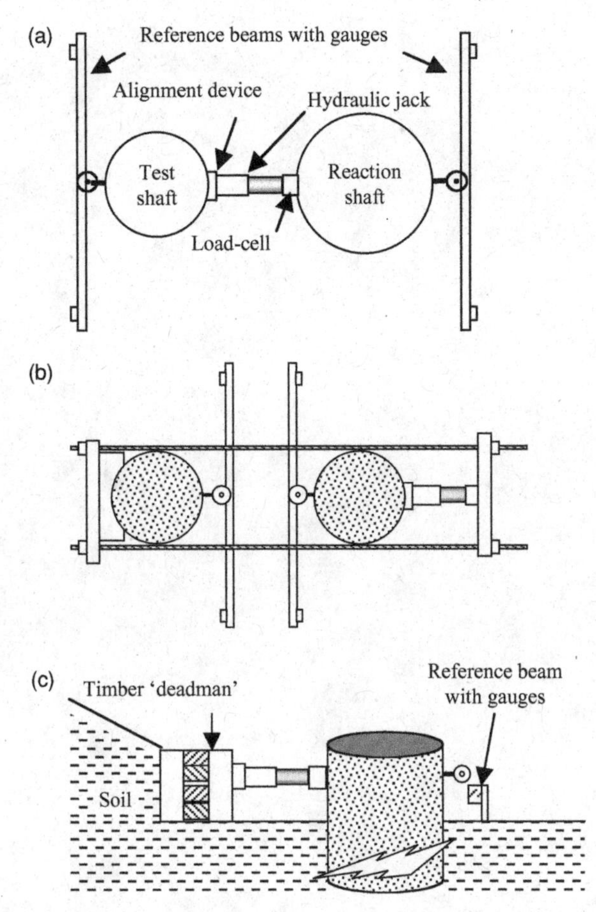

Figure 5.7 (a) Plan-view of a two-shaft lateral load-test arrangement. (b) Plan-view of an alternative lateral load-test arrangement. (c) Lateral load-test arrangement for a single-shaft using a timber 'deadman'

Where a single shaft must be tested, a reaction mass can be constructed by installing a timber 'deadman' against the side of an excavation or soil stockpile (Figure 5.7(c)). If site topography does not lend itself to a soil-supported deadman, one possible alternative arrangement is to construct a kentledge 'on grade'.

In some cases, the actual shape of the shaft when laterally loaded is important, since it shows where the maximum bending takes place. In these cases, in addition to the reference beams and dial gauges or LVDTs, the test shaft is instrumented with inclinometer tubes so that inclinometer measurements can be made at various stages of the loading cycle. A procedure for setting up and performing lateral load tests is given in ASTM D3966 – 'Standard Test Method for Piles under Lateral Loads'. Other alternative designs for a reaction mass are also given in D3966.

6

High-strain Testing for Capacity and/or Integrity

6.1 HIGH-STRAIN DYNAMIC (DROP-WEIGHT) TESTING OF DRIVEN PILES

High-strain testing is generally divided into two categories – static or 'pseudo-static' load testing, which usually measures actual foundation capacity after installation or construction, and high-strain dynamic testing (HSDT), which is most commonly used to monitor the installation of driven piles and predict their load capacity on completion. The idea of predicting capacity from observations made during pile driving has been around for more than a hundred years. A.M. Wellington proposed such a formula in 1892. Wellington's formula was adopted by the *Engineering News*, and became known as the 'Engineering News formula' (Wellington, 1892). With some modification, the HSDT method can also be used to measure the load capacity of drilled shafts or ACIP piles and verify shaft integrity.

Typically, HSDT is performed by impacting the top of the shaft with a mass that has sufficient energy to either drive the pile a short distance into the ground or mobilize the end-bearing capacity of the shaft. For capacity determination, HSDT may be applied in one of two ways, similar to static loading:

- Proof testing, in which the stress developed in the shaft is equivalent to twice the design load, similar to a static-load proof test.
- Ultimate capacity, in which the stress developed in the shaft is sufficient to shear all side friction and cause a permanent displacement of the toe of the shaft that exceeds the maximum settlement allowed in the design, similar to a 'static-load test to failure'.

Nondestructive Testing of Deep Foundations B.H. Hertlein and A.G. Davis

For integrity testing purposes, much less energy is required and it is usually possible to verify the length and integrity of a shaft without physically driving it into the ground. However, since pile driving momentarily shears the side friction on the shaft, there is much less attenuation of the stress waves and the integrity of long slender shafts can be verified in soil conditions where the Impulse-Echo or Impulse-Response methods, which are discussed later in this book, would be inconclusive.

The amount of energy imparted to the pile by the hammer impact is determined by the mass and shape of the hammer, the drop height and the type of 'cushion' that is used to protect the top of the pile. The impact of the mass may last up to about 100 ms, depending on the amount of energy required. The impact generates stress waves that propagate down the pile and are partially reflected back by any zone where material properties vary significantly, or from the end of the shaft. This is the same principle as that employed by the low-strain hand-held hammer techniques discussed later in this book, but the strain caused by the stress waves generated by the pile-driving hammer is several orders of magnitude higher than the low-strain techniques; hence, the descriptive term 'high-strain' tests.

The origins of the high-strain technique are difficult to trace with certainty, but it is well-documented that E.A.L. Smith of the Raymond Pile Driving Company developed the first practical numerical method for pile-driving analysis in the 1950s (Smith, 1960). What is less well-known is that Smith built on the work of David Isaacs, an Australian engineer who realized in the 1930s that the techniques and formulae that worked so well on timber piles would be fraught with problems if applied directly to the concrete piles that were just then being developed. Isaacs apparently recognized the potential problems of tensile waves in concrete piles, and the problems inherent in 'over-driving' (Isaacs, 1931). The stress wave generated by the hammer impact propagates down the pile as a compression wave. If the tip of the pile is in relatively soft soil, it is effectively a 'free' end and the compression wave changes to a tension wave when it reaches the pile tip. In cases of 'over-driving', the pile is struck too hard and the energy imparted by the hammer creates such a high tensile strain that the concrete cracks. Figures 6.1(a) and 6.1(b) are simple schematics that illustrate the basic physics involved.

Smith's work was taken up by researchers at Texas A&M University, where L.L. Lowery and T.J. Hirsh developed a program that took advantage of the computing power that was just beginning to become available in the late 1960s. Lowery and Hirsh reviewed the work that had already been done (Lowery *et al.*, 1969) and then set about examining and improving many of the assumptions made in Smith's model (Hirsh *et al.*, 1976). The results were published under the aegis of the Texas Transportation Institute (TTI) and are still in use today in some parts of the world. This model, however, was based on simple steam or air hammers and does not address the complexities introduced by the combustion cycle of the diesel hammer. That was addressed in the work of Goble at Case Institute of Technology.

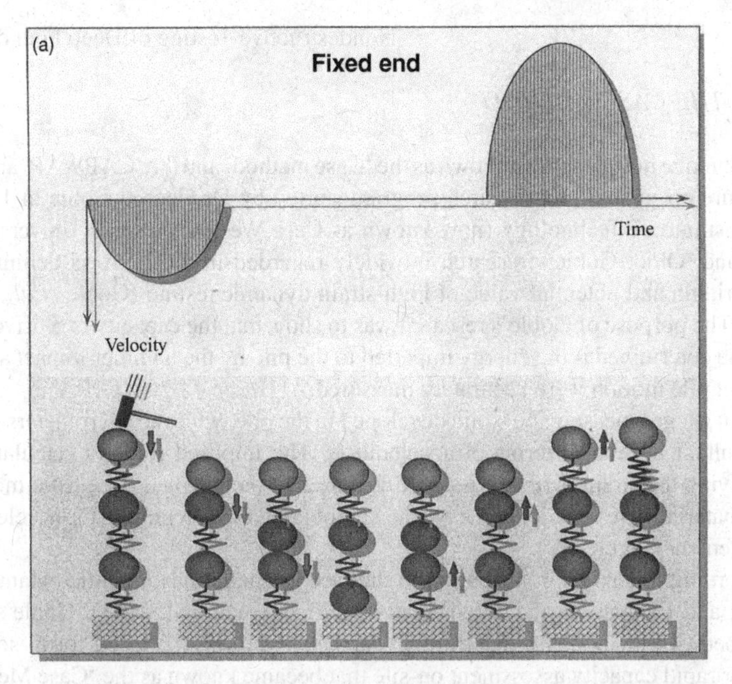

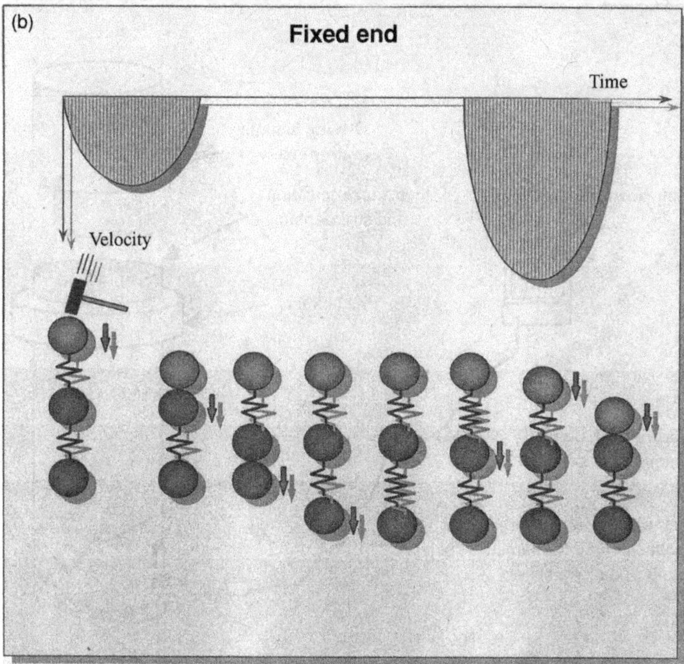

Figure 6.1 Schematics of elastic wave propagation in (a) a shaft with a fixed end (stiff soil or rock at base) and (b) a shaft with a free end (soft soil conditions). Reproduced by permission of Profound BV, The Netherlands

6.1.1 THE CASE METHOD

The technique now generally known as the 'Case method' and the 'CAPWAP' analysis procedure originated in a research program started by Dr George Goble in 1964 at Case Institute of Technology (now known as Case Western Reserve University) in Cleveland, Ohio. Goble's research is widely regarded in the USA as defining the basic criteria and potential value of high-strain dynamic testing (Goble *et al.*, 1975, 1980). The purpose of Goble's research was to show that the capacity of a driven pile could be determined if the energy imparted to the pile by the hammer impact and the resultant pile motion were accurately measured.

Strain gauges measure the strain developed in the pile, while accelerometers record the resultant motion in terms of acceleration. The imposed force is calculated by multiplying the strain by the cross-sectional area of the pile and the elastic modulus of the material. The velocity of the pile head is obtained by integrating the acceleration measurement (Figure 6.2).

According to Likins' contribution to the Deep Foundations Institute 'Manual for Testing and Evaluation of Drilled Shafts' (DFI International, 2003), Goble's team developed two methods for determining capacity. One was a 'closed-form' solution used for rapid capacity assessment on-site that became known as the 'Case Method',

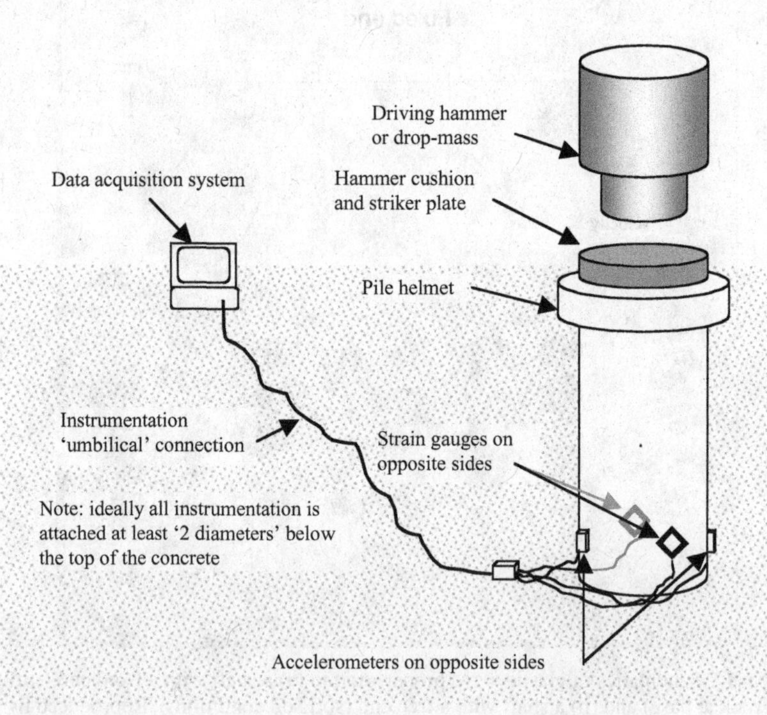

Figure 6.2 Schematic of high-strain test instrumentation

after the university. The other was a variation of the already established wave equation analysis (Smith, 1960). Goble's team replaced the wave-equation hammer model with force and velocity measurements. A mathematical model of the pile is created, using assumed soil parameters. The measured velocity is then used as a boundary condition for the pile head and the force required to keep the system in dynamic equilibrium is computed. The computed force is then compared with the force calculated from the strain measurements. If the two do not match, the soil parameters are modified and the calculation is repeated until a good agreement is reached between the measured force and the computed force. The technique therefore determines the soil behavior from the pile-driving measurements and became known as the Case Pile Wave Analysis Program or 'CAPWAP' (Rausche *et al.*, 1972).

6.1.2 *THE TNO METHOD*

The Dutch team, led by Henk van Koten, expanded on the work of De Josselin de Jong, who first examined what happens in the soil during pile driving in a report published in *De Ingenieur (The Engineer)* in 1956 (De Josselin de Jong, 1956) and Verduin, who reported on stress-wave measurements in piles in a TNO report in 1956 (Verduin, 1956). The Dutch were thus working along similar lines to the USA team (van Koten, 1967) and reached a similar solution – the TNO-Wave Analysis method.

The resulting pile and soil models from either analysis program can then be processed to generate a simulated static load-test result. When sufficient time has passed after pile installation to allow soil strength changes to stabilize, using this signal-matching technique on data recorded from a 're-strike' of the shaft that produces a minimum of 2.5 mm (0.1 in) permanent set per blow usually produces close agreement with static load-test results (Likins *et al.*, 1996).

Key factors for an efficient pile-driving program are the hammer weight and the drop height. The correct combination of these two must be selected if the pile is to be driven efficiently, while damage from excessive stress is to be avoided. A major benefit of high-strain dynamic testing is that the energy input to the pile is measured accurately. The energy input is obtained from the work done on the pile as the product of the measured force and velocity integrated over time. Since the shaft cross-section is known, the stresses imposed by driving can be calculated. Prior to such instrumentation, hammer size and drop height could only be calculated from theoretical models, which made no realistic provision for energy losses caused by friction, misalignment or distortion of the hammer 'cushion'.

High-strain testing was also found to be useful in detecting damage caused by improper handling or driving of the piles (Rausche and Goble, 1979). The method is often used to evaluate the integrity of piles where unusual driving behavior has been observed, such as advancing more rapidly or driving deeper than expected.

In small projects, with small numbers of piles, it is usual to perform dynamic testing on only one or two. For sites with larger numbers of piles, such as a silo or tank base, it is common practice to install the first few production piles in several different locations across the site and perform dynamic tests on each one as a check on site variability and installation consistency. Where several hundred or even thousands of driven piles will be required, a pre-contract test pile program will enable the foundation design to be refined and perhaps reduce the number or length of piles required. Random dynamic tests during pile installation can then be compared with the test program results as a quality assurance measure. Such a pre-contract test program may also enable the engineer to reduce the number of static load tests required for the site.

The first standard for the test method was published by the American Society for Testing and Materials (ASTM) in 1986, as D4945 'High Strain Dynamic Testing of Piles'. The subsequent widespread acceptance of high-strain testing is demonstrated by other national standards of practice and its regular appearance in standard specifications issued by State and Local Government agencies in many countries.

From the pile-driving contractor's point of view, stress-wave theory comes into the picture long before the job starts. Choosing the appropriate equipment for the job is not simply about using a bigger hammer on bigger piles. The soil conditions will have a significant effect on the rate at which the pile penetrates for each blow of the hammer and therefore the stresses created in the pile during driving. If the hammer is too large for the job, and is used incautiously, it can create such high tensile forces in the pile that the concrete cracks. If the hammer is too small for the job, then it will be inefficient at driving the piles, so costing the contractor both time and money.

In addition to measuring shaft integrity and predicting capacity, stress-wave theory can be used to predict driving performance of a given hammer/pile combination in specific soil conditions. The program originally developed by Goble *et al.* (1980) was known as the 'Wave Equation Analysis of Piles' and given the acronym WEAP. A more recent version by the same company is known as GRLWEAP. This program simulates driving conditions to predict hammer performance, driving stresses and 'set per blow'. TNO developed a similar product, known as the 'Pile-Driving Prediction' program or PDP-Wave. In either case, the choice of hammer type and selection of hammer weight and stroke can be assessed numerically before actually mobilizing equipment to the site. In fact, many contractors base their bid estimate for a project on predictions of the time required to drive each pile, as calculated by the GRLWEAP, PDP-Wave or a similar program.

For the reader who wishes to learn more about the various types of pile-driving hammers and the factors that affect their efficiency and performance, the DFI International publication 'Pile Inspector's Guide to Hammers' presents a very concise summary (DFI International, 1995).

6.1.3 THE EFFECT OF SOIL AND OTHER FACTORS

In addition to the proprietary hardware systems and analysis methods, numerous other formulae and methods have evolved for analysis of high-strain test data, each having its own particular characteristics that affect the accuracy and consistency of capacity predictions. Research conducted by Professor James Long at the University of Illinois at Urbana compared six of the most commonly used formulae, and found all but one to be in reasonable agreement when used under optimum conditions (Long *et al.*, 1999).

The six formulae that Long compared were:

(1) EN – *the Engineering News* formula that was developed by Wellington (Wellington, 1892).
(2) GATES – the Gates formula, published in 1957 in *Civil Engineering* (Gates, 1957).
(3) WEAP – the Wave Equation Analysis Program (Goble and Rausche, 1986).
(4) PDA – the Pile Driving Analyzer (Hannigan, 1990).
(5) ME – the Measured Energy Method (Paikowsky *et al.*, 1994).
(6) CAPWAP – the Case Pile Wave Analysis Program.

Long *et al.* provide outline descriptions of each method in their 1999 paper. These descriptions are summarized here. The EN and the GATES formulae are purely mathematical formulae, based on empirical observations. They share weaknesses in common with other simple dynamic formulae in that they consider only the kinetic energy of driving, assume constant soil resistance regardless of penetration velocity and ignore the length and axial stiffness of the pile.

The WEAP analysis can be used to predict pile capacity and hammer performance. The choice of hammer for a particular job is often based on the results of a preliminary WEAP analysis. It has also been established that the accuracy of pile-capacity estimates is improved if the actual energy delivered to the pile is measured in the field and used as an input into the analysis.

The PDA method requires measurement of the pile-head acceleration and the strain developed in the pile shaft during driving. Acceleration is converted to penetration velocity and strain is converted to force. The Case method then uses these inputs to predict pile capacity. There are several versions of the Case method, each of which produces a slightly different prediction. Hannigan's paper reviews them in more detail (Hannigan, 1990).

The 'Measured Energy' approach uses essentially the same set-up as the PDA, with recordings of acceleration and strain, but adds direct measurement of pile set during driving. Like the PDA method, the ME approach can be used to predict capacity in the field during the driving process.

CAPWAP combines data from the PDA method with a refined WEAP analysis. Using the recorded acceleration history to predict force, the method then compares predicted force with measured force. If the two do not match, the input parameters for the soil are modified and the calculation is repeated. This iterative approach makes the method impractical for site use and it is more commonly performed in an office after driving is completed.

Figure 6.3(a), taken from Long *et al.* (1999), shows the comparison of capacity measured in static load tests with predictions from the various methods, using data

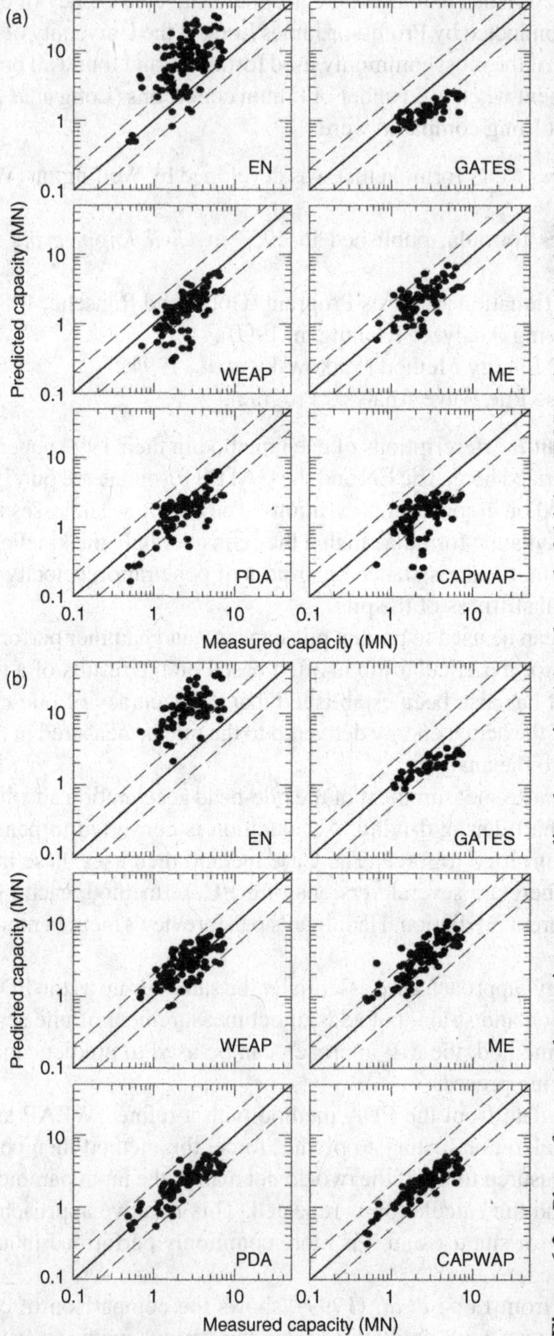

Figure 6.3 Plots of measured capacity versus predicted capacity: (a) at the end of driving (EOD); (b) at the beginning of 're-strike' (BOR) (after Long *et al.*, 1999). Long, J.H., D. Bozhurt, J.A. Kerrigan and M.H. Wysockey, Value of Methods for Predicting Axial Pile Capacity. In *Transportation Research Record 1663*, Transportation Research Board, National Research Council, Washington, D.C., 1999, Figures 1 and 2, page 59. Reproduced with permission of TRB

recorded at the end of driving. Not surprisingly, the more modern methods, with sophisticated instrumentation, fared better than the earlier, purely empirical methods. There is, however, no room for complacency, even with the most sophisticated measurement systems. Figure 6.3(b) shows the same comparison for data recorded at the beginning of 're-strike', after the soil has had time to recover from the disturbance of driving. Clearly, soil 'set-up' has a significant influence on the accuracy of dynamic capacity predictions.

6.2 HIGH-STRAIN TESTING OF DRILLED SHAFTS AND AUGERED, CAST-IN-PLACE PILES

According to publications by Pile Dynamics, Inc., high-strain dynamic testing of drilled shafts in North America was performed in 1974 on a barrette for a steel mill in Mexico and in 1977 for a housing project in Charleston, West Virginia, where numerous ACIP (augered, cast-in-place) piles were tested. Following a dynamic test on a drilled shaft for the Sunshine Skyway Bridge replacement in the early 1980s, a project was conducted in Australia to reduce the extensive amount of static testing that had been planned (Rausche and Seidel, 1984; Seidel and Rausche, 1984). These early tests were conducted using modified pile-driving equipment and the data processed in much the same way as for driven piles. The problem with this approach is that there is no unique solution to the capacity calculation and the analyst's experience is a crucial factor in the accuracy of the result. The high-strain dynamic test works well on a shaft of uniform section and material quality, such as a steel pile or a high quality pre-cast concrete pile, but the inherently variable cross-section and material quality of a drilled shaft or an ACIP pile introduce variables that can complicate the calculation of capacity to an unrealistic degree.

6.2.1 CEBTP SIMBAT

In Europe, the CEBTP (Centre Experimental de Recherche et d'Études du Bâtiment et des Travaux Publics) developed a somewhat different approach to high-strain testing, specifically for the drilled shaft and augered, cast-in-place pile applications.

In 1979, the CEBTP was sponsored by its parent organization, the French Construction Industry Federation, to develop a high-strain dynamic test applicable to bored piles (drilled shafts). The result of this research was a methodology known as SIMBAT. The name is derived from the phrase 'Simulation de Battage' (Driving Simulation) (Heritier and Paquet, 1986; Paquet, 1987, 1988; Heritier *et al.*, 1991). While this method has gained acceptance in France, Eire and the UK and to a lesser extent, in Italy and Spain, it is, at the time of writing this book, rarely seen elsewhere.

Similarly to the dynamic models used for other high-strain methods, SIMBAT is based on the propagation of waves in long, elastic cylinders. When the shaft head is struck with a falling weight, the resulting stress waves travel down the shaft to the toe, where they are reflected back to the shaft top. In a free, undamped shaft, the particle velocity and amplitude of the return waves would be similar to the original waves. When the pile is surrounded and restrained by soil, part of the wave is reflected back up at each and every external restraint, with the remainder of the wave continuing downward. Therefore at any moment, there are both upward and downward forces and velocities in the shaft.

The SIMBAT technique is able to separate these upward and downward forces, $F\uparrow$ and $F\downarrow$, and then calculate the dynamic soil reaction, R_{dy} as the difference between the upward force in a free pile and the real measured upward force. The conversion of dynamic to static reaction (R_{stat}) is carried out by expressing R_{dy} as:

$$R_{dy}(z, V_{pen}) = R_{stat}(z) f(V_{pen}) \tag{6.1}$$

where z is the cumulated shaft penetration and V_{pen} is the velocity of shaft penetration into the soil; $f(V_{pen})$ is obtained from the SIMBAT set of test results by a regression method.

An integral part of the procedure is numerical simulation, where the experimental signals are introduced into the program and compared with theoretically generated signals. The choice of a realistic pile/soil model is critical and the model used in SIMBAT is shown in Figure 6.4. This is a model with two degrees of freedom, as opposed to the classical (CASE) spring–dashpot with a single degree of freedom. To exploit this model, varying relative penetration velocities have to be induced into the

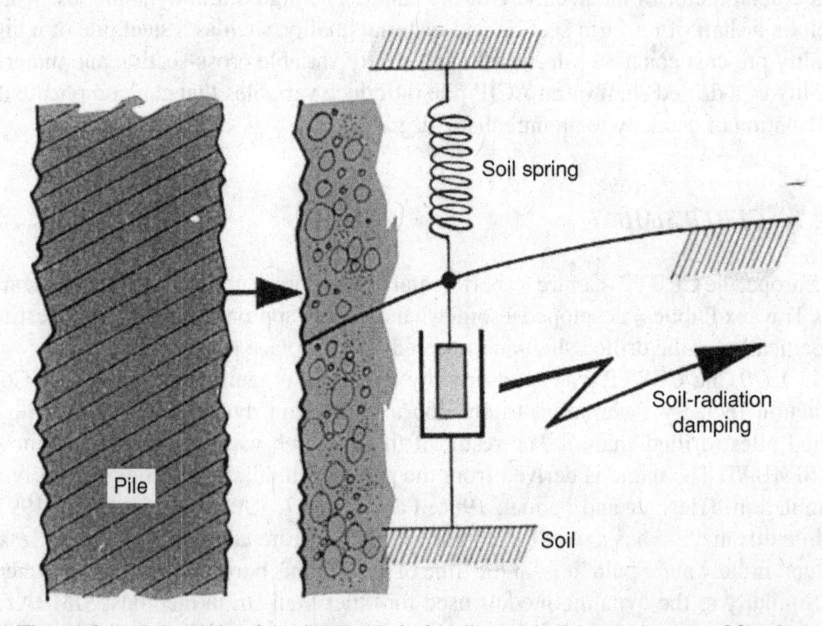

Figure 6.4 Schematic of the SIMBAT shaft/soil model with two degrees of freedom

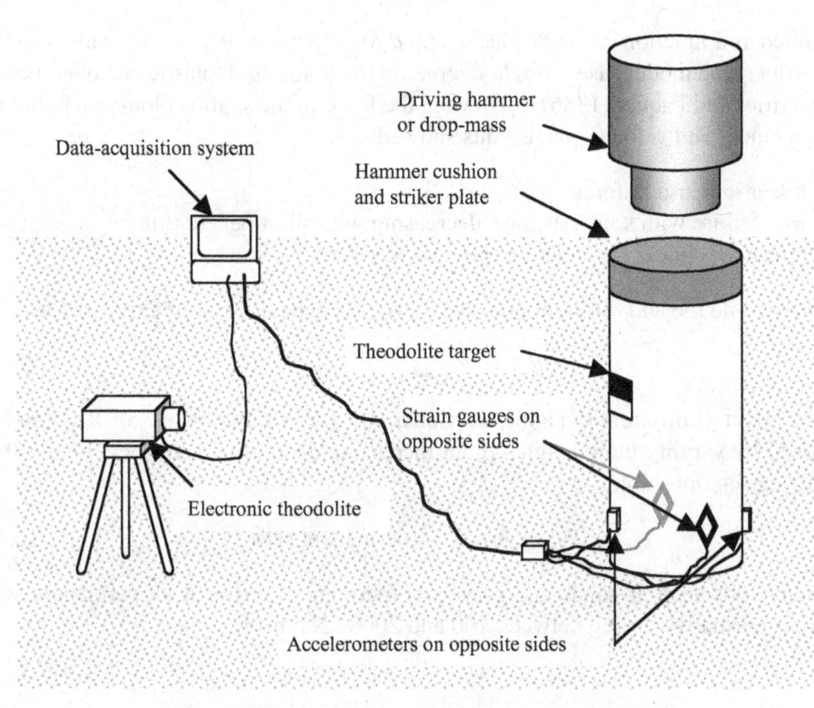

Figure 6.5 Schematic of SIMBAT instrumentation

pile during the test. In addition, an accurate method of measuring both temporary shaft compression and permanent set is required. The SIMBAT method uses an electronic theodolite placed about 3 to 5 m from the pile head to achieve this (Figure 6.5).

A single free-fall drop-hammer is used and the hammer drop height is progressively increased/decreased to obtain the variable V_{pen} needed to satisfy the SIMBAT model requirements.

Heritier and Paquet (1986) described work on instrumented piles that showed that the force (F) at the pile toe at any time (t) could be expressed as follows:

$$F(t) = a\gamma(t) + bv(t) + cd(t) \tag{6.2}$$

where γ is the acceleration, V is the velocity, d is the diplacement, and a, b and c are all variables associated with the soils, such as damping, viscosity and elasticity.

This relationship was demonstrated by tests on a pile with accelerometers fitted at a small distance above the pile toe. Very low amplitude blows were applied to the pile head, in order to remain as much as possible in the elastic domain. The pile/soil impedance for each blow can then be obtained by Fourier transform on the time-base signals, as in low-strain stress-wave testing. These observations showed that the dynamic soil resistance is a function of both the shaft velocity and its displacement – the resistance, $F(v)$ increases more rapidly than $F(d)$. However, if the total force, F, is plotted as a function of the displacement, d, then F rises very rapidly and it can be seen why some models use a single-degree-of-freedom, rigid–plastic behavior. From the Heritier and Paquet (1986) tests, when the force in the shaft is plotted in both the displacement and velocity planes, this showed:

- A linear increase in force.
- Brittle failure with soil resistance decreasing with shaft penetration.
- A return to a linear regime after three wave pulses.

For the pile toe and different pile segments, this behavior can be expressed as:

$$R_{dy} = R_{spring} + R_{damping} \tag{6.3}$$

The effect of different dynamic reactions, R_{dy}, for different values of V_{pen} can be assessed by varying blow heights in an increasing/decreasing sequence, giving the following relationship:

$$R_{dy} = R_{stat} + f(\text{permanent set}) \tag{6.4}$$

In many soils such as sands and gravels, $f(\text{permanent set}) = K_c \times \text{permanent set}$, with K_c normally varying between 100 and 2000 kN mm^{-1}.

6.2.2 *SIMBAT TEST METHODOLOGY*

An extension to the shaft head is built as described in Section 6.2 above. This extension is required to be at least '2.5 shaft diameters' high and 'well-reinforced' against bursting under the drop-weight impact. As this cap is to serve as a dynamometer, it must be smooth, cylindrical and of good quality, homogeneous concrete. Two strain gauges, two accelerometers and an electronic theodolite target are mounted on the side of the cap, '2 diameters' below the top. Then, a series of impacts are made, with the hammer drop height being progressively increased and decreased, as shown in Figure 6.6 (Stain and Davis, 1989).

The test signals are processed as follows:

- The measured accelerations and velocities are corrected using the direct theodolite measured displacements.
- The upward and downward forces $F\uparrow$ and $F\downarrow$ are separated (Figure 6.7).
- Measurement of the dynamic soil reaction, R_{dy}.
- Regression analysis to obtain $f(V_{pen})$ and thus K, the 'dynamic-to-static factor'.
- The elastic shaft compression is calculated.
- A plot of total shaft settlement versus load is generated (Figure 6.8).

Finally, a set of computer simulations are made both to verify the measured signals and to calculate the distribution of soil resistance down the shaft and at the pile toe.

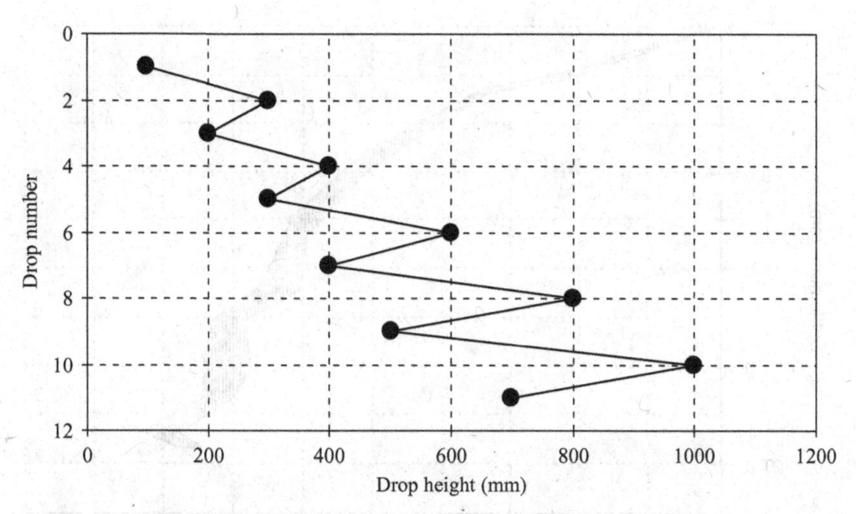

Figure 6.6 Typical series of drop heights for the SIMBAT test

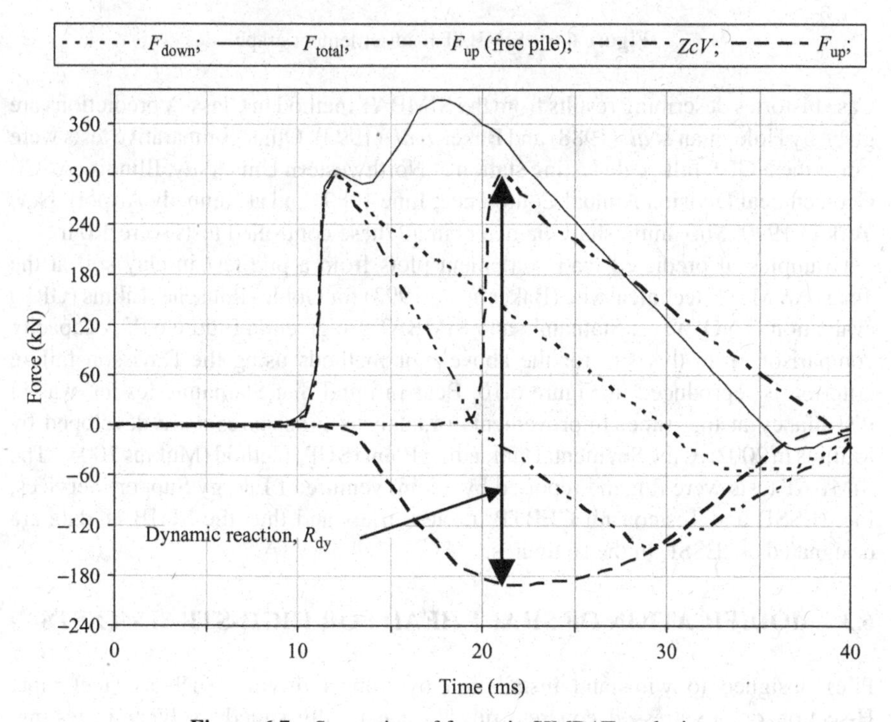

Figure 6.7 Separation of forces in SIMBAT analysis

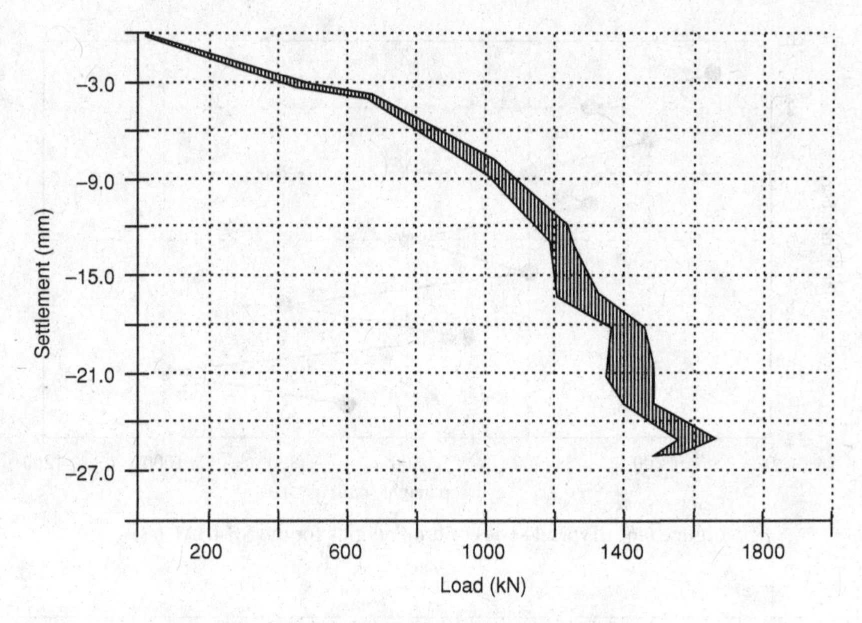

Figure 6.8 SIMBAT load/settlement graph

Case histories describing results from the SIMBAT method in Class-A predictions are given by Holeyman *et al.* (1988) and Baker *et al.* (1993). Other comparative tests were run at the NGES full-scale testing station at Northwestern University, Illinois (ASCE Geotechnical Division Annual Conference, June 1989) and at Kennedy Airport, New York in 1990. Maximum shaft diameters in all these controlled tests were 1.0 m.

Examples of predicted load–settlement plots from a pile test in clay soil at the Texas A&M Geotechnical Site (Baker *et al.*, 1993) for Goble–Rausche–Likins (GRL) evaluation, TNO-Wave, Statnamic and SIMBAT are given in Figure 6.9. A capacity comparison from this site for the above four methods using the Davisson failure criterion is reproduced in Figure 6.10. Bear in mind that Statnamic testing was in it's infancy at that time. Improvements in analysis procedures were developed by Mullins in 2002; as the Segmental Unloading Point (SUP) method (Mukins 2004). The SIMBAT tests were run and reported by a joint venture of Energy Support Services, Inc. (ESSI) and Testconsult CEBTP in these trials and thus the SIMBAT data are designated as 'ESSI' in the se figures.

6.3 MODIFICATION OF SHAFT HEAD FOR HIGH-STRAIN TESTS

Piles designed to withstand installation by impact driving, such as steel pipe, H-section or pre-stressed concrete piles, are naturally suited to dynamic testing. Drilled shafts and ACIP piles, however, are not designed to accommodate the high stresses of drop-weight impacts and must therefore be modified to avoid damage to

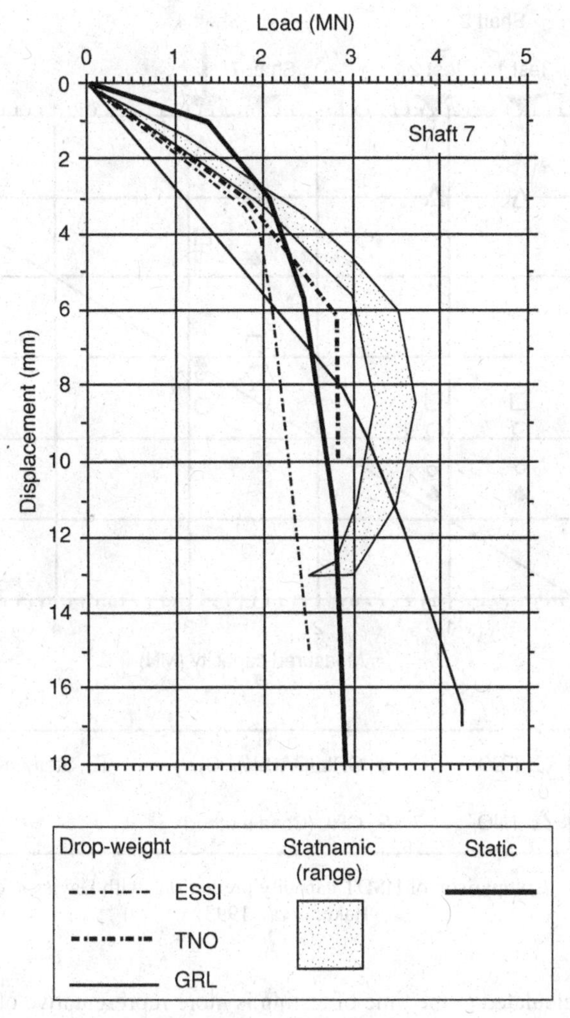

Figure 6.9 Predicted load/settlement plots from an FhWA research project (after Baker *et al.*, 1993)

the shaft head when high-strain dynamic testing is to be used. Typically, extra helical reinforcing steel is incorporated into the upper few feet of the shaft to help the concrete withstand the stresses caused by the high-strain test.

In addition to the structural modifications, the concrete in the shaft must be allowed to gain sufficient strength to withstand the stress of testing. In most cases, this means allowing the concrete to mature seven days or so before testing. Longer periods, up to forteen days or more, are sometimes preferred to allow additional strength gain. This delay also allows the soil to recover from the disturbance caused by drilling so that

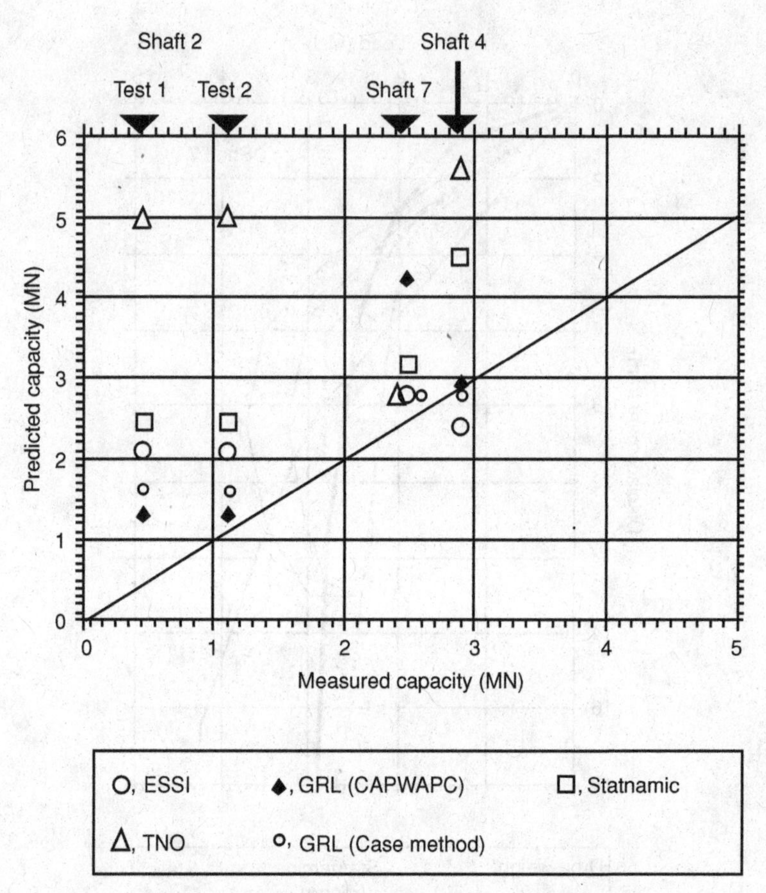

Figure 6.10 Comparison of HSDT capacity predictions with Davisson criteria (after Baker *et al.*, 1993)

the capacity calculated at the time of testing is more representative of the long-term behavior of the shaft/soil complex.

One of the modifications necessary for performing high-strain tests on drilled shafts is to protect the reinforcing steel. Most drilled shafts and ACIP piles have reinforcing steel extending from the top of the shaft for connection to the reinforcing steel in the column, 'grade-beams' or pile caps. The typical solution is to cast a short concrete extension to the pile to enclose the reinforcing steel. The extension can be formed in a short length of steel casing, in which case additional helical steel is usually not needed, or in a timber or cardboard form, in which case additional helical or hoop steel will be required to resist the impact stresses. On completion of the testing, the built-up portion of the shaft is removed to expose the reinforcing steel for connection to the superstructure.

The top of the shaft extension should be troweled smooth and perpendicular to the longitudinal axis of the shaft so that a minimum of packing material is needed to achieve uniform distribution of energy from the drop-weight or hammer into the shaft. For maximum accuracy of test interpretation, it is preferred that the packing material be limited to a single layer of plywood or similar material that distributes the energy of the hammer or drop-weight impact uniformly over the cross-section of the shaft.

An advantage of the pile-top extension is that it eliminates the need for excavation to attach the test transducers. Typically, the strain gauges and accelerometers are attached about '2 diameters' below the pile head to ensure that impact stresses are evenly distributed in the shaft at the measurement point. Part of the casing or formwork will need to be removed or have 'windows' cut in it to allow access for installation of the transducers.

In the event that it is not practical to build up the shaft by '2 diameters', the transducers may be attached closer to the top of the shaft, although the impact stresses may not be evenly distributed at that level. The recommended practice in such a situation is to use four transducers instead of the usual two, equally spaced around the shaft perimeter. Recent variations of the technique have used an accelerometer on the drop-weight to measure deceleration force and a radar gun to measure drop-weight velocity (Robinson *et al.*, 2002). Optical techniques, such as laser velocity measurements, are also being experimented with. The purpose of these variants is to reduce the complexity of the instrumentation that must be affixed to the side of the shaft.

6.4 PRACTICAL CONSIDERATIONS FOR DROP-WEIGHT TECHNIQUES

It is now generally accepted that the mass of the drop-weight or hammer for high-strain testing of drilled shafts should be at least 1.5 to 2 % of the capacity to be proven (Hussein *et al.*, 1996). The use of smaller masses will require larger drop heights to generate higher impact velocities, which may result in unacceptably high stresses within the shaft.

Typical drop-weights are relatively unsophisticated, being pieces of I-beams welded together or sections of steel casing filled with concrete or steel scrap. The most important feature of each weight is the ability to deliver its kinetic energy efficiently and accurately to the top of the shaft being tested. As long as it can accomplish this consistently, its appearance is usually of little importance! In order to ensure consistent accuracy of the contact between the drop-weight and the shaft, a set of guides or 'leads' is used to direct the drop-weight.

Drop mechanisms vary. On older cranes, the drop-weight may be lifted by crane cable and dropped by releasing the cable-drum brake. Most modern cranes, however, cannot release the cable drum in this manner, and must use 'power down' to drop the weight. This does not give a true 'free-fall condition', which complicates the calculation of impact energy. Several alternative drop methods have been developed in response to this problem. The most common is a cam or latch release mechanism, in which the drop-weight is raised by crane or winch until it is held by latches built into the leads. The lift cable is disconnected or spooled out to free the drop-weight. The latches or cams are then withdrawn to release the drop-weight.

The initial drop height is usually fairly low to enable the operators to assess impact concentricity, drop-weight/shaft alignment and uniformity of energy distribution, and adjust the drop-weight leads accordingly. When alignment is satisfactory, the Case and TNO methods increase drop heights until either the shaft settlement per blow is at least 2.5 mm, or until the required capacity is exceeded or the shaft fails. The CEBTP method alternates drop heights in a high–low pattern to develop a more complete picture of the shaft's response over the loading range.

Regardless of which proprietary algorithm is used to calculate the mobilized capacity in a high-strain test, the limitations are similar. If the set per blow is less than about 2.5 mm, the test results will be very conservative, largely because the lower part and the toe of the shaft have not moved and end-bearing has not been mobilized. Thus, the calculated capacity will be based purely on side friction for the upper part of the shaft and will therefore be conservative. It follows, then, that a more accurate estimate of capacity will be derived from higher-energy impacts, either from higher-drop heights or larger-drop-weight masses.

A brief analysis is made after each impact to determine if the calculated capacity is sufficient or if additional, higher-energy drops are required. The accuracy of this analysis is partly dependent on certain assumptions or estimates of shaft shape. If the soils are such that a regular cross-section can be assumed (such as stiff clays) or if the approximate cross-section has been estimated from installation monitors, such as those commonly used on ACIP pile rigs in Europe (and becoming accepted in the USA at the time of writing), then the time required for the analysis procedure is significantly reduced. The shaft shape, however, has more effect on the load distribution throughout the shaft than on the actual capacity.

6.4.1 NEWTON'S APPLE

A relatively recent innovation by the testing firm GRL Engineers has got around the need to excavate the top of the shaft by making the force measurement on the ram. GRL designed a completely integrated high-strain testing system specifically for use on drilled shafts and augered, cast-in-place piles. They say that they named the system 'Newton's Apple' because it really is smart! The force sensor is claimed to provide greater accuracy than calculation of force from strain measurements when concrete quality in the upper portion of the shaft is variable or in doubt (Rausche and Robinson, 2000).

Newton's Apple has a square steel support frame and ram guide. The ram is modular and its mass can be varied over the range 5–20 tons which allows the generation of dynamic loads up to 2000 tons. The system can be configured as a free-standing unit, with the ram supported by a latch on the external frame until dropped or the ram can be dropped directly by the crane if the crane is equipped with a free-fall clutch and can be set up to control the back-lash or 'whip' of the boom when the ram is released.

With a footprint of 1.8 m × 1.8 m, and a height of 6 m, Newton's Apple can be handled on site by a mobile crane, making set-up and operation relatively fast compared with the older high-strain drop-mass systems. In a demonstration at the National Geotechnical Experimental site in Amherst, MA, USA, GRL unloaded the equipment off a truck, conducted high-strain tests on three separate shafts and loaded everything back on the truck in less than 7 h (Rausche and Robinson, 2000) (Figure 6.11).

Since Newton's Apple has eliminated the need for strain gauges, it has also eliminated the need for excavating or building-up the shaft if adequate reinforcing is incorporated into the head of the shaft during initial construction.

6.5 HSDT ALTERNATIVES

The need to reinforce or modify the upper portion of the shaft has always been considered problematic in some situations and researchers other than GRL found that it was more advantageous to modify the dynamic test instead. A high-strain test in which the same load is applied but is built-up over a greater time period significantly reduces the stresses developed in the top of the shaft and eliminates the need for special reinforcing in the shaft head.

6.5.1 THE STATNAMIC METHOD

In 1989, Patrick Bermingham of the Berminghammer Corporation, Ontario, Canada, published his invention of the mechanical components of the 'Statnamic' method, which is a direct practical application of Newton's second and third laws of motion:

- Force equals mass times acceleration.
- For every action, there is an equal and opposite reaction.

The Statnamic method uses solid propellant fuel to launch a reaction mass upwards and thus, by reaction, develops a downward load on the shaft. The original data-analysis procedures for the method were developed by Peter Middendorp at The Netherlands Organization for Building Construction and Research (TNO).

Bengt Fellenius gave a little of the history behind this simple-sounding description in his opening address at the First International Statnamic Seminar (Fellenius, 1995). Fellenius had been searching for someone to construct a drop hammer to his

Figure 6.11 Typical 'Newton's Apple' set-up for a medium-sized shaft. Reproduced by permission of GRL Engineers, Inc., Ohio, USA

specification but with virtually no budget. Patrick Bermingham expressed his interest, and, after some consideration, suggested the idea that the load be lifted, taking advantage of Newton's laws, rather than dropped. At first, the idea was little more than a mechanical novelty but a chance encounter between Patrick Bermingham and Peter Middendorp at a deep foundations conference quickly turned into a collaboration that resulted in the development of the Statnamic method.

The basic mechanism is relatively simple. The reaction mass is modular, consisting of steel segments that mount on a mandrel, allowing the reaction mass to be tailored to the load required. The mandrel is mounted atop a cylinder that fits over a piston mounted on the top of the pile. The piston assembly includes a load-cell and a laser displacement transducer. The fuel charge is placed in the cylinder. The mandrel and reaction mass are supported within a frame that includes a catch system to stop the reaction mass from 'crashing back' down onto the shaft after the propellant has been fired. The rapidity and duration of loading can be controlled by varying the amount of fuel and the rate of 'burn'. The Statnamic method can therefore achieve quasi-static load durations, as discussed by several contributors to the 2nd International Statnamic Conference (Kusakabe *et al.*, 2000).

The essence of the method is that the duration of the loading period is longer than the natural period of the foundation shaft. This means that the shaft remains in compression throughout the measurement period and thus there are no reflected stress waves to contend with in the resulting measurement data. The shaft is instrumented with strain gauges and accelerometers. The data from these, coupled with the data from the load-cell and the displacement transducer in the Statnamic equipment, provide a very complete picture of the response of the shaft.

The analysis of Statnamic data, however, although simpler than for the high-strain dynamic tests, poses its own set of challenges. At first glance, Newton's second and third laws of motion appear to deal with this problem. The problem is that the measured Statnamic force must be corrected for inertia and damping effects before the capacity of the foundation can be reached. The original solution, developed by Peter Middendorp, served to win acceptance for the method and turn it into a viable alternative to static load testing. Middendorp proposed that the only unknown, the damping factor, could be calculated by using a procedure known as the 'Unloading Point Method' (UPM) (Middendorp *et al.*, 1992). Calculations using the UPM, however, only approximated the behavior of a shaft under static load. While the results were close enough to win acceptance for this method, it took advances in computing power and modeling procedures to lift the method to the next level. In 1999, Professor Gray Mullins at the University of South Florida published the 'Modified Unloading Point' (MUP) and the 'Segmental Unloading Point' (SUP) methods. These methods consider the distribution of acceleration throughout the shaft by recognizing that concrete is an elastic medium (Mullins *et al.*, 2002). The result of Mullins' work has been a significant improvement in the correlation between static load tests and Statnamic test results. Illustrations of typical Statnamic set-ups for a medium-sized shaft and

Figure 6.12 Typical Statnamic set-up for a medium-sized shaft. Reproduced by permission of Applied Foundation Testing, Inc., Florida, USA

a large-diameter shaft, plus a comparison of Statnamic and a traditional 'Kentledge' for the same-sized shafts, are shown in Figures 6.12–6.14, respectively.

As with most test methods, the Statnamic test offers its own unique advantages. Because it is propelled by a controlled fuel-burning system rather than being dropped and relying on gravity, the Statnamic system can be laid 'on its side' on a suitable arrangement of skids or runners to perform a lateral load test (Figures 6.15 and 6.16). This discovery was a significant milestone in the development of the Statnamic test. In 1998, Professor Dan Brown of Auburn University, Alabama, USA, published the results of some experiments in lateral load testing with the Statnamic system (Brown, 1998). In his analysis, the spring stiffness and viscous damping of the system are varied until a good match is achieved between the theoretical curve and the measured

Figure 6.13 Typical Statnamic set-up for a large-diameter shaft. Reproduced by permission of Applied Foundation Testing, Inc., Florida, USA

Figure 6.14 Comparison of Statnamic and a traditional kentledge for the same-sized shafts. Reproduced by permission of Applied Foundation Testing, Inc., Florida, USA

Figure 6.15 Lateral Statnamic test on a small-diameter shaft using skids. Reproduced by permission of Applied Foundation Testing, Inc., Florida, USA

Figure 6.16 Lateral Statnamic test on a large-diameter shaft using a barge. Reproduced by permission of Applied Foundation Testing, Inc., Florida, USA

Figure 6.17 Statnamic test over water. Reproduced by permission of Applied Foundation Testing, Inc., Florida, USA

data. The results can then be used to create a computer model for the rest of the site, using well-established modeling software, such as LPILE or FLPIER.

In addition, because the loading device rises as the test is performed, instead of dropping, as in the other dynamic methods, the test can be readily adapted to 'over-water' applications (Figure 6.17).

The ability to control the rate and duration of loading enables the results of the Statnamic test to resemble the effects of static loading more closely than the high-strain dynamic tests – hence the name!

6.5.2 THE FUNDEX METHOD

The importance of the ability to provide a longer-duration controlled loading rate was recognized at about the same time in Europe and led to the development of the Fundex PLT. The Fundex Company of Belgium developed their own rapid-load test method, which essentially uses a bed of springs to cushion the impact from a larger drop-weight, thus spreading the loading period over more time than either the Case or TNO methods. Fundex worked with American Piledriving, Inc., of Pleasanton, California, to build a pile-load test (PLT) rig (Figure 6.18) to service the US West-Coast market. A high-profile project that compared the results of static and Fundex load test results at the University of California in Berkeley was presented to the 2002 DFI conference in San Diego (Presten and Kasali, 2002).

Large coil springs are fixed to the underside of the drop-weight and an anvil plate with a load-cell is set on the top of the pile. The springs distribute the impact of the hammer evenly over the pile head and extend it over time, so that the rate of stress rise in the shaft is significantly reduced. A typical impact will last about 400 ms, about 200 ms of which are effective in loading the pile (Figure 6.19).

A hydraulic system built into the drop-weight guide frame (or leads) catches the drop-weight as it 'bounces off' the test shaft, thus preventing a second impact. The load-cell measures the force and an optical displacement transducer measures the movement of the pile head. The optical transmitter is attached to the head of the shaft, while and the receiver is set on the ground about 15 to 20 m away. This distance is enough to ensure that the accuracy of the measurement is not affected by instability, caused by the Rayleigh or surface vibration waves which radiate from the shaft after the impact of the loading mass.

A significant advantage of the Fundex method is that the entire loading assembly is mounted on a crawler crane or tracked carrier and can thus be rapidly moved around the site, so allowing multiple tests to be performed in a single day. A schematic of the Fundex pile-loader tester is shown in Figure 6.20.

There is not, at the time of writing, any standard for the performance of the Statnamic or Fundex tests, since the high-strain test standard does not apply to the lower-strain rate developed by these methods and the response of the shaft/soil complex is substantially different. However, the ASTM Subcommittee D18-11, 'Deep Foundations', is working on the development of a standard, under the designation Work Item WK219 – 'test Method for Piles Under Rapid Axial Compressive Force'. The Fundex and Statnamic methods have become known generically as variants of 'The Rapid Load Test'. The Testing and Evaluation Committee of the Deep Foundations Institute is also currently preparing a set of practical guidelines for the application, performance and analysis of the 'Rapid Load Test'.

Regardless of which dynamic test is considered, the cost will be considerably less than for performing static load tests if several shafts are to be tested. The equipment for all of the dynamic techniques is readily transportable and can be set up in a few hours, unlike the process of building a stable kentledge or installing reaction shafts, which can take several days. In addition, the dynamic techniques can be applied

Figure 6.18 The Fundex pile-load test (PLT) rig. Reproduced by permission of American Piledriving, Inc., California, USA

relatively easily in conditions that would be very difficult or impossible for static loading methods, such as on individual shafts in deep water.

6.6 LIMITATIONS OF HIGH-STRAIN DYNAMIC TESTING

The preparation of the shaft head described earlier can be a limitation for the high-strain dynamic test, since the concrete used to build-up the shaft must reach a suitable strength before the test can be applied. Even if the shaft is designed and constructed to withstand the stresses caused by dynamic testing, the concrete will still require a period of curing in order to gain sufficient strength. This factor is somewhat less of a limitation for the 'Rapid Load Test' methods.

Figure 6.19 The Fundex pile-load test (PLT) drop-mass showing the multiple springs. Reproduced by permission of American Piledriving, Inc., California, USA

Full end-bearing must be developed in order to accurately predict the total capacity of the shaft. Tests in which only a small displacement of the toe is achieved will yield conservative results. The amount of 'under-prediction' will depend on the stiffness of the bearing stratum at the shaft toe.

Like most nondestructive test methods, high-strain dynamic tests do not measure the required property directly. They measure related properties and by application of 'basic physics', the operator interprets the data to yield the required information. The accuracy of the test is therefore very much dependent on the skill and experience of both the tester and the analyst and on the validity of the assumptions made during the application of the test, and in the interpretation of the resulting data.

It is also important that the effect of soil properties that change with time is not underestimated when using the high-strain analysis techniques. Despite the relatively long history of these methods, there is still some controversy within the industry about the factors that affect the accuracy of the capacity predictions made with high-strain methods. The research performed by Professor James Long on the effects of time on pile capacity has already been discussed earlier in this chapter. Long analyzed data from tests performed at the end of driving (EOD), when the soil-pore pressure is highest but side friction has been sheared, and at the beginning of 're-strike' (BOR) on the same shafts a few days later, when sufficient time has elapsed for the temporary increase in pore pressure to dissipate and the side friction has been re-established – a process known as 'set-up' (Long *et al.*, 1999). Work described in a Keynote Lecture

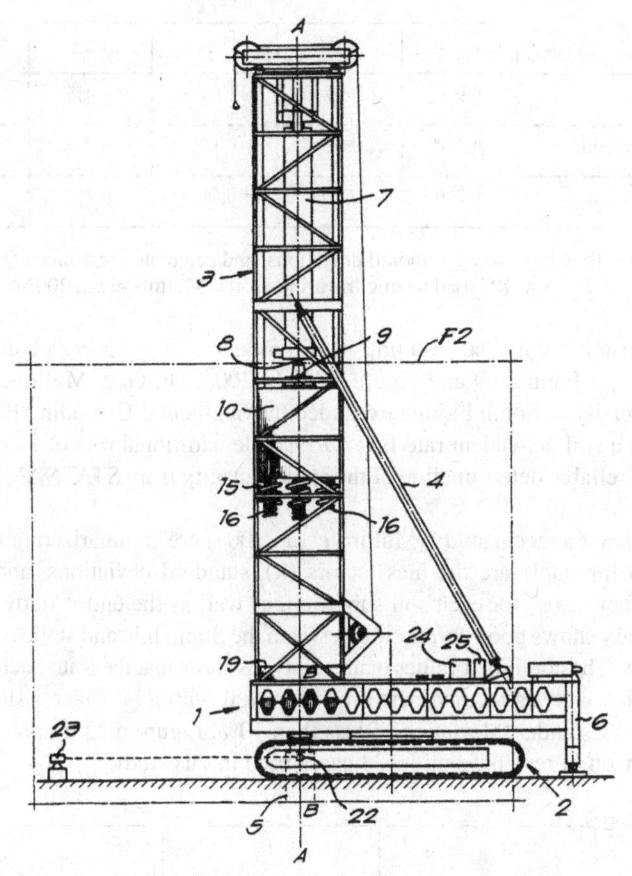

Figure 6.20 Schematic of the Fundex pile-load tester from US Patent Application 5 325 702 (1994). Reproduced by permission of American Piledriving, Inc., California, USA

by S.J. Paikowsky at the Stress Wave 2000 Conference (Paikowsky and Stenersen, 2000) culminated in the publication of a load-test database for the FHwA, in which capacity predictions are compared for piles tested at the end of driving.

Paikowsky's work showed that the capacity of shafts in some soils increased as the 'soil set-up', whereas in other soils the dissipation of pore pressure led to relaxation of the soil and a drop in shaft capacity, in which case the dynamic tests would tend to 'over-predict' shaft capacity, thus reducing the safety factor for the unwary engineer. In general, it appears that capacity predictions for driven piles are most accurate when using data from 're-strike tests', after soil conditions have stabilized somewhat. It is, therefore, reasonable to assume that a similar increase in accuracy can be expected for tests on ACIP piles and drilled shafts if sufficient time is allowed between the end of construction and application of the tests.

Historically, a limitation of the STATNAMIC method has been that its analysis procedure was a moving target, constantly changing and evolving. A variety of methods

	Rock	Sand	Silt	Clay	All Soils
Bias Factor, λ	0.999	0.994	1.041	1.035	1.017
Standard Deviation	0.068	0.083	0.116	0.119	0.097
Resistance Factor, φ	0.739	0.726	0.737	0.730	0.734

Figure 6.21 Bias factors (λ), standard deviations, and calculated resistance factors (φ) for STATNAMIC load testing in various soils (Mullins *et al.*, 2004)

have been used in the past, resulting in inconsistent and poor correlations, such as those shown in Figure 6.9 and Figure 6.10. In 2003, Dr. Gray Mullins, Ph.D., P.E. of the University of South Florida expanded the Segmental Unloading Point Method (SUP) with a soil dependent rate factor to handle additional rate of loading effects. As a result, reliable determination of the static capacity from STATNAMIC testing is now at our disposal.

The results of a recent study (Mullins *et al.*, 2004) are summarized in Figure 6.21. Included in this table are the bias factors (λ), standard deviations, and calculated resistance factors (φ) for each soil subgroup, as well as the entire study (All Soils). This summary shows good correlation between the Statnamic and static capacities for all soil types. The combined values of all soil types show that the Rate Factor corrected SUP capacity analysis method performs very well with a bias factor slightly larger than 1.0 and a standard deviation of less than 10%. Figure 6.22 graphically depicts the comparison of results from the 34 cases used in this study.

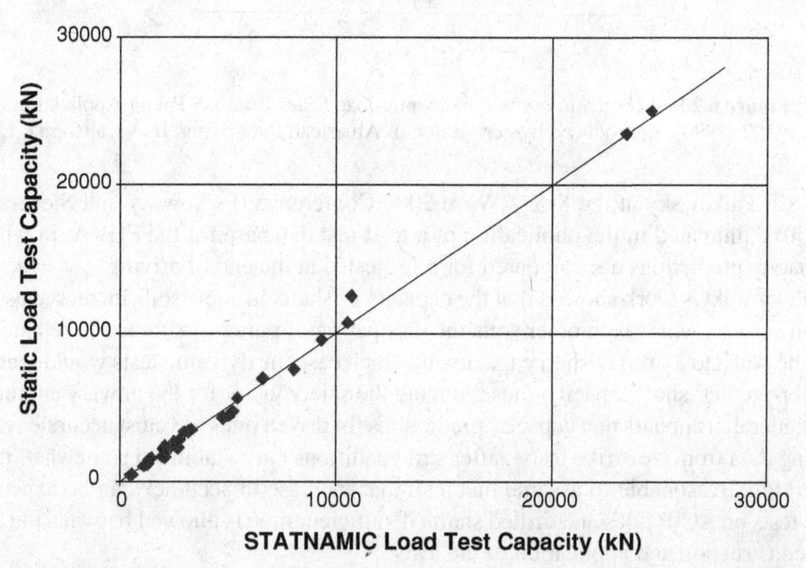

Figure 6.22 Comparison of 34 Statnamic and static load tests (Mullins *et al.*, 2004)

7

Low-strain Surface Tests – Sonic Echo

Although appearing nearly 50 years ago, low-strain pile integrity testing experienced a very slow early commercial development. At the beginning of the 21st Century, however, these techniques figure among the most classical NDT methods in Civil Engineering. Great strides in the fields of data acquisition under sometimes-difficult site conditions, miniaturization of test equipment and the introduction of digital signal processing techniques are the principal reasons for this.

However, while it is now relatively easy to obtain high-quality test signals, the physical problem of pile/soil interaction has not changed. For this reason, the deep foundations community sanctions two main families of low-strain test methods at the present time:

- Shaft head impact tests (Sonic Echo and Sonic Mobility), where the response to an impact on the head of the pile shaft is measured by a transducer coupled to that shaft head.
- Cross-hole or down-hole tests (Sonic Logging, Gamma–Gamma Logging and Parallel Seismic) where pre-placed tubes in or adjacent to the shaft act as guides for sensors.

In the former approach, the response signal is conditioned by the shaft–soil interaction, whereas the latter measures the shaft material conditions only. A further distinction can be made between the two groups, in that damping of the response signal by the soil around the pile shaft in the former tests results in reduced information being received with increasing depth, whereas the latter tests are independent of the shaft depth. As a result, head-impact tests require specialist interpretation with

Nondestructive Testing of Deep Foundations B.H. Hertlein and A.G. Davis

understanding of their limitations, while cross-hole and down-hole tests are specified by agencies requiring routine quality control testing of shafts depending on uniform concrete quality at and just above the shaft base.

7.1 SONIC ECHO (IMPULSE ECHO)

This test first surfaced in the literature in 1968 as 'la méthode d'écho' or 'echo method'. The test has received several names since then, leading to some confusion, and has variously been referred to as seismic, seismic reflection, sonic, echo, PIT, TNO-Wave and various others. It must not be confused, however, with another stress-wave variant, the Pulse-Echo method, which is an *ultrasonic* pulse velocity test typically used for assessing the integrity of metal structural elements.

The 'echo method' started in 1968 with the publication of Jean Paquet's paper on nondestructive testing of piles in the *French National Building and Civil Engineering Annals* (English translation by Xiang Yee (Yee, 1991)). Needless to say, the test equipment used at that time was limited to analog technology (we didn't know that digital techniques were possible at that time!) and data storage was a dream for the future. This first paper dealt with the application of electric transmission line theory to the one-dimensional problem of compression stress bar-waves transmitted from the pile head down the shaft. Paquet explored the limitations of the method, particularly with respect to excitation frequency bandwidth. Bar-wave analysis in both time- and frequency-domains was discussed (see Appendix I).

Paquet laid down the fundamental theories for stress-wave transmission down piles, including changes in pile and soil dynamic impedance and damping of stress waves by pile material and surrounding soil. However, he considered the Sonic-Echo principle to be simple! He compared the method to ultrasound techniques used for metal testing, stressing the difference that metal ultrasonic testing uses a stream of waves containing tens of periods of sinusoids with well-defined frequencies, with great lateral dimensions of the emitter in relation to the emitted wavelength and a consequent directional effect. This method encounters several difficulties when applied to piles:

- A significant damping of higher frequencies in concrete piles caused by the unhomogenous nature of the concrete.
- Wave propagation damping caused by the lateral soil.
- Difficulty in coupling a directional emitter to the pile head.

Moreover, practical conditions at the worksite have to be allowed for (cleanliness and soundness of pile heads, control testing considered to be of secondary importance on construction sites, etc.). The theory considers the pile as a straight-sided prism. Often, the actuality is a pile head appearing as a badly defined concrete mass, sometimes covered by water and/or mud and with projecting steel bar reinforcement. Extrapolation of theoretical and laboratory techniques to these conditions is not guaranteed.

Since transmission conditions in concrete rendered impossible the wave-train approach, Paquet settled for an impulsive excitation force with a very wide frequency band, such as a hammer blow. All practitioners of the Sonic-Echo method have adopted such an approach to this present day. The inconvenience of such a signal is that it cannot be filtered at the pile-head return and that it cannot be transmitted directionally. The third difficulty of coupling of the emitter to the pile head means that great care on site is required to ensure no debonding or cracking exists in the concrete immediately below the receiving sensor location. Ideally, a series of receivers (four or more) placed at different points around the impact point would eliminate some of the problems, such as surface waves at impact confusing the received signal.

Paquet also introduced the notions of deconvolution of the return signal to eliminate the oscillatory return to equilibrium of the pile, as well as exponential amplification of the time-based response to reduce the soil damping effect and enhance signals from deeper reflectors. These principles are applied today in modern testing equipment.

The predominant concrete pile construction method in France in 1970 was the bored cast-*in situ* pile (known variously in different parts of the world as caisson, drilled shaft or cast-in-place pile). Paquet (1968) and Briard (1970) showed that for bored piles with the analog signal processing technology available at the time, the best approach was to analyze the vibration response in the frequency-domain, particularly when defects in the top few meters were to be detected. At this point, a divergence occurred between the frequency-domain approach developed in France and applied in the UK (predominantly bored-pile construction) and the approach adopted in Holland, the Scandinavian countries and the USA where pre-cast driven piles were more common and the Sonic-Echo time-domain analysis was adequate for piles with constant cross-sections (Weltman, 1977; Rausche and Goble, 1979; van Koten *et al.*, 1980). Early attempts to apply the Sonic-Echo method to drilled shafts in the USA were limited and advocated the installation of sensors down the body of the shaft to help interpretation (Baker and Kahn, 1971; Steinbach, 1971; Hearne *et al.*, 1981).

Major developments of the Sonic-Echo method in the 1980s were motivated by the need for equipment and interpretation techniques to produce fast and reliable results from every type of piling site. Several companies and research organizations developed, manufactured and sold their own proprietary systems, such as the CEBTP in France with the MIMP (from Methode IMPulsionelle), Pile Dynamics, Inc. in the USA with the Pile Integrity Tester (PIT) and TNO in Holland with the echo test. In the last decade, several additional companies have developed their own equipment, significantly broadening the choice of supplier for the practitioner.

The original Sonic-Echo method was the basis for the development of other methods, such as 'Vibration', 'Impulse Response', and 'Impedance Log'. Any improvement in signal acquisition, analysis and processing in the Sonic-Echo method has a direct bearing on improvements in these techniques. A schematic of the basic Impulse/Echo-Test set-up is presented in Figure 7.1.

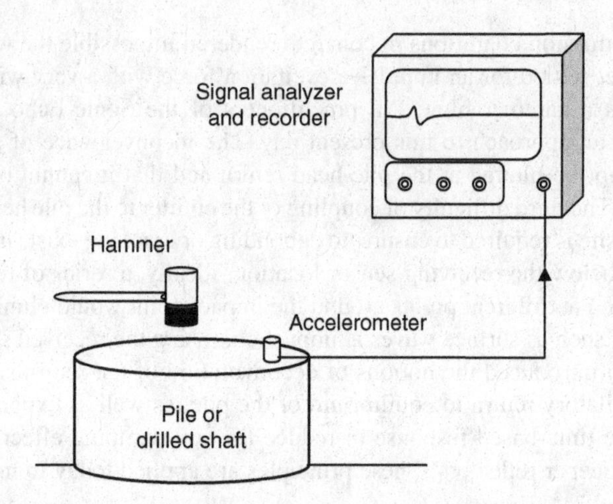

Figure 7.1 Schematic of the basic Impulse/Echo-test set-up

7.1.1 TEST PRINCIPLE

The Sonic-Echo test is performed by striking the pile head with a light hammer and measuring the response of the pile with a sensor (accelerometer or geophone velocity transducer) coupled to the pile head. The hammer blow generates a compressive stress wave which is channeled down the pile shaft as a 'bar-wave'. The latter is partly reflected back towards the pile head by any change in impedance within the pile. These impedance changes can be as a result of changes in pile section, concrete density or shaft–soil properties. The stress wave is transmitted through the pile at a velocity, v_b (where v_b is the bar-wave velocity of propagation through the pile material) and the time lapse, t, between the hammer impulse and the arrival of the reflected waves at the pile head from pile tip is a measure of the distance traveled by the stress wave, such that:

$$t = 2L/v_b \quad (7.1)$$

where L represents the distance to the reflecting surface (pile tip in this case).

If the value of v_b is known, or can be estimated within reasonable limits, then t will give an estimate of the pile length or the depth to any other reflecting surface within the pile. If the pile length is known, then a comparison can be made between the length calculated from the test result and the known length, in order to verify that the depth to the reflecting surface is correct.

7.1.2 TYPICAL TEST PROCEDURE

The material in the pile head must be prepared such that no delamination or 'micro-cracking' is present, to ensure a clean transmission of the stress wave down the pile.

The sensor is coupled to the pile head, usually with a grease- or gel-based couplant, and the pile is struck with the hammer at or near the pile axis. Normally, a hammer weighing less than 1 kg with a plastic impact tip is used. Heavier hammers have sometimes been found to give better results for large piles greater than 1 m in diameter (see Figure 7.1).

The test is repeated several (at least three) times in order to obtain representative samples by averaging of these individual results. The more hammer blows recorded, then the greater the reduction in the effects of random signals (noise) from other site activities or system noise. As the effects of this extraneous noise are reduced, so the repeatable parts of the signal are enhanced. As a general rule, background noise can be reduced by a factor of $\sqrt{n}$, where n is the number of superimposed signals from tests on the same pile.

The testing procedure has been standardized by the American Society for Testing and Materials (ASTM) in ASTM D5882, 'Standard Test Method for Low Strain Integrity Testing of Piles', and is referenced in ACI Report 228.2R (1998).

7.1.3 DATA PROCESSING AND DISPLAY

The signal is usually processed to give a display of the sensor velocity (or displacement) as a function of time. For most systems, a sampling rate of 10 kHz, with a time array length of approximately 100 ms, is suitable to encompass all pile lengths normally encountered. If an accelerometer is used as the sensor, then integration of the accelerometer signal is required to obtain velocity. A typical velocity–time response plot is shown in Figure 7.2. This plot is referred to as the signal–response curve and

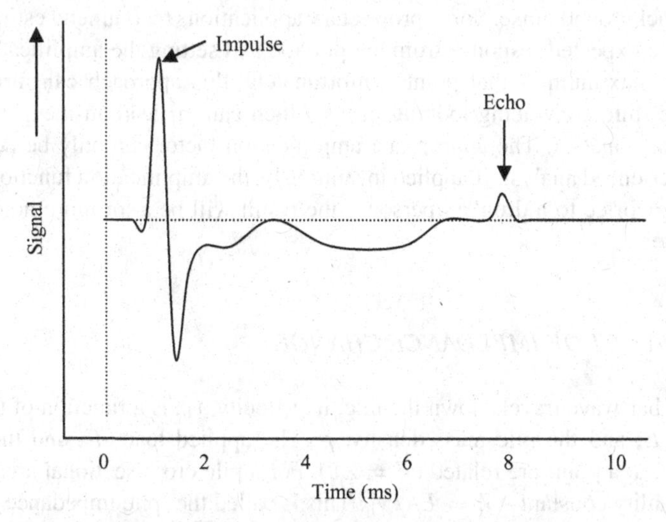

Figure 7.2 Typical example of a Sonic-Echo velocity (signal)–time response plot

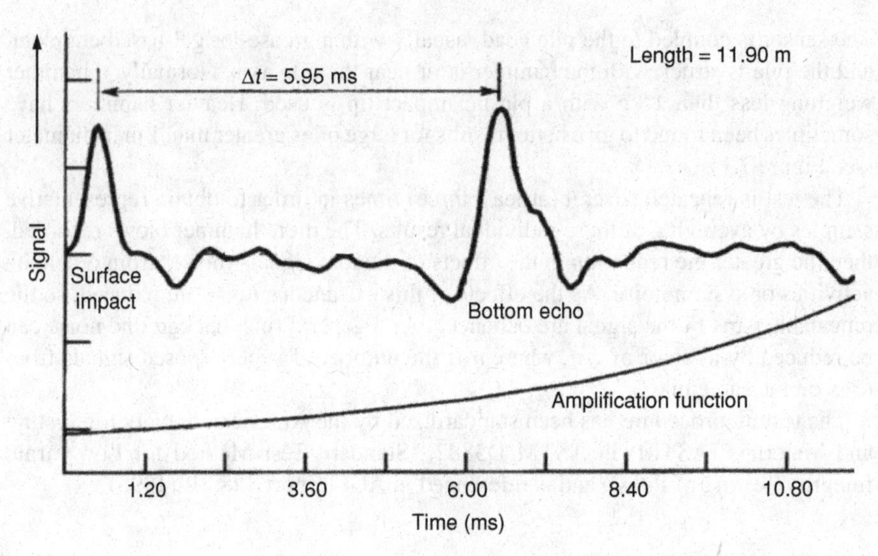

Figure 7.3 Amplified Sonic-Echo velocity (signal)–time response plot

can also be displayed as velocity versus pile depth by converting time to depth using equation (7.1). Typical values for bar-wave velocities in concrete piles range from 3700 to 4300 m s^{-1}.

Because the attenuation of the returned signal increases with the distance of travel, it is usually beneficial to exponentially amplify the signal, increasing with time, in order to identify weak reflections from the vicinity of the pile toe, as shown in Figure 7.3. Great care must be exercised in applying this function, however, since it will also amplify background noise. Some proprietary applications recommend estimating the time for the expected response from the pile toe and setting the amplification factor to be at a maximum at that point. Unfortunately, this approach can turn random 'geo-noise' into convincing-looking peaks which can mislead an inexperienced or over-anxious analyst. The appropriate amplification factor can only be determined by an experienced analyst. If applied incautiously, the amplification function is much like giving coffee to a drunken person – the result will be a lot more noise, but no more sense!

7.1.4 EFFECT OF IMPEDANCE CHANGE

When the bar-wave travels down the pile, its velocity, v_b, is a function of the elastic modulus, E, and the pile mass density, ρ. The applied load, F, and the particle velocity, v, at a point are related ($F = Zv$). For a pile cross-sectional area, A – the proportionality constant – $Z = EA/V_b$. This is called the 'pile impedance', because it is a measure of the pile's resistance to velocity.

Suppose that, at some depth down the pile shaft, the impedance changes from Z_1 to Z_2. When the downwards-traveling stress wave, F_i, arrives at this point, part of the wave is reflected upwards (F_u) and part transmitted downwards (F_d), such that both continuity and equilibrium are satisfied. Solving the simultaneous equations gives:

$$F_d = F_i[2Z_2/(Z_2 + Z_1)] \tag{7.2a}$$

$$F_u = F_i[(Z_2 - Z_1)/(Z_2 + Z_1)] \tag{7.2b}$$

For a uniform pile, $Z_2 = Z_1$ and neither F_d nor F_u are generated and the input wave continues unchanged. An extreme example is the pile toe (free end) where Z_2 is zero. The compressive downward wave will be completely reflected upwards and the resulting F_u will be of opposite sign (tensile). A decrease in either area, A, or modulus, E, produces a tensile upwards reflection (compressive in the event of an increase in A or E). This means that a tensile reflected wave (such as from the pile toe) arriving at the surface produces a positive 'blip' or peak on the signal–response curve, while a compressive reflection from an increase in A or E produces a negative 'blip' on the response curve. A compressive upwards-traveling wave can also be generated by an increase in the lateral soil resistance.

In this way, distinction may be made between possible breaks/cross-section losses in the shaft and increases in pile cross-section ('bulbs') and/or soil resistance. However, the magnitude of the impedance change controls the sensor signal amplitude of that impedance change at the pile head. Ellway (1987) suggested that the most sensitive test equipment at that time was capable of detecting pile impedance changes of approximately 1:0.8 (or 1:1.2). These low levels of impedance change allow most of the signal to continue downwards through the feature. As the ratio approaches 1:0.5 (or 1:2), most of the wave is reflected, so causing a 'clear echo'. When the ratio change exceeds 1:0.25 (or 1:4), Ellway also suggested that the incident stress wave is almost completely reflected, with no further information being obtained from below this point.

Signal-matching techniques were introduced to exploit these impedance distinctions given by the Sonic-Echo method (van Koten and Wood, 1987; Middendorp and Reiding, 1988; Rausche *et al.*, 1988). In these techniques, the pile is divided into a series of continuous segments and the wave reflections and transmissions are computed at each segment boundary. The 'real-pile' signals can then be matched to the simulated signals by iterations with segment shapes and sizes down the pile shaft, as well as varying soil resistances. Catalogs of typical signal traces have been developed by different testing groups in order to facilitate signal-matching on site. Typical examples of such signal-matching results are given in Figure 7.4.

The biggest single difficulty facing most testers on site is that of convincing the contractors to provide suitable access to perform the test, and to prepare the shaft head appropriately to ensure accurate and valid test results. The scenarios shown in Figures 7.5–7.7 are, unfortunately, distressingly common sights for foundation shaft integrity testers all over the world!

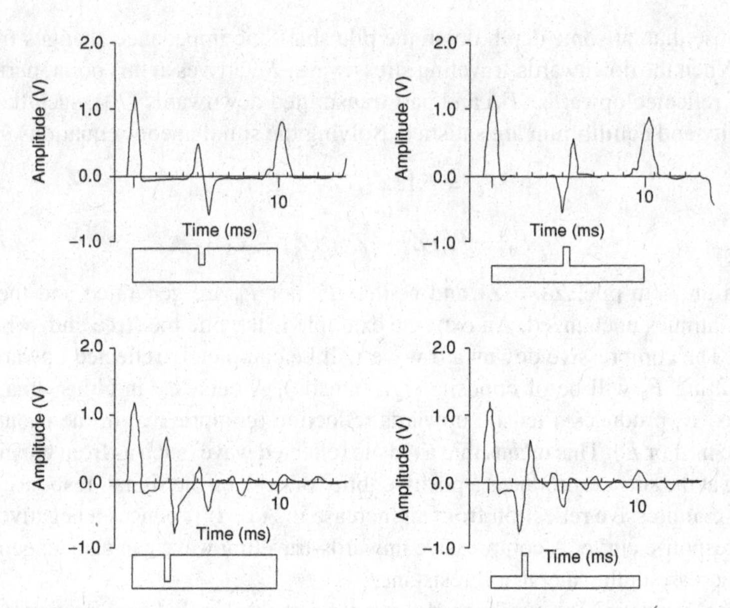

Figure 7.4 Examples of typical signal/shape-matching traces (after Rausche *et al.*, 1998)

Figure 7.5 Shaft head inaccessible for testing due to a cap-reinforcing cage

Figure 7.6 Shaft head inaccessible for testing due to inadequate excavation

Figure 7.7 Shaft damaged by site equipment and inadequate excavation

7.1.5 USE OF MULTIPLE RESPONSE TRANSDUCERS – DOUBLE SENSOR TESTING

High-speed data acquisition, made economically practical by the relatively recent introduction of analog-to-digital converters capable of sampling rates greater than 100 kHz per channel in multi-channel mode, has opened up new possibilities for existing test methods. Sometimes, it is required to measure the length and to evaluate the integrity of piles under existing structures. The nondestructive test methods commonly used for the evaluation of free piles and concrete drilled shafts, such as Impulse Echo and Impulse Response, are not always applicable to the problem of testing older piles under existing structures, because:

- Free access to the pile head is required in order to perform the test.
- The pile-length calculation requires an assumption of the stress-wave velocity in the pile material.

Most concrete piles have bar-wave velocities, v_b, of 3900 to 4100 m/s and errors in length determination are not significant. In these authors' experience, older piles can have v_b values ranging from 3300 to 4100 m/s, depending on the type and condition of the concrete in the pile. This value can also be modified by the presence of permanent steel casings, such as tube piles and 'Raymond step-taper piles'. Additionally, steel piles typically have v_b values ranging from 5000 to 5800 m/s. It is therefore essential to know v_b if reliable predictions are to be made of the lengths of existing piles.

A combination of two methods is normally used to test existing piles (Davis, 1995):

- Sonic Echo
- Parallel Seismic (PS).

In certain cases, pile length can be determined accurately by a modification to the Sonic-Echo method alone. A recently developed approach to testing older piles using two sensors attached to the pile shaft just below the pile cap is described here.

The Sonic-Echo method uses a small impact delivered at the head of the pile shaft and measures the time taken for the stress wave generated to travel down the shaft and to be reflected back to a receiving transducer (usually an accelerometer or velocity transducer) coupled to the pile head. Both the moment of impact and the pile-head vertical movement are recorded on a time-base. If the response from the pile tip is clear and v_b is known, then the pile length is obtained directly from equation (7.1), given earlier in this chapter.

The energy from the hammer blow is small and the damping effect from the soils surrounding the shaft will progressively dissipate that energy as the stress wave travels up and down the shaft. To increase information from the test, the signal response is often progressively amplified with time, as discussed earlier in this chapter.

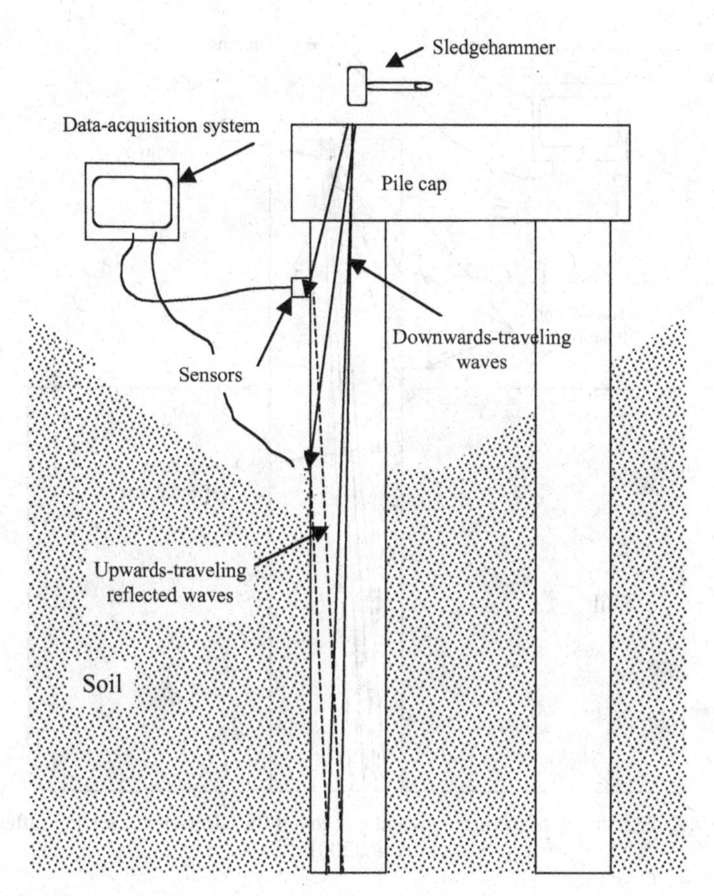

Figure 7.8 Schematic of the arrangement employed for dual-sensor Impulse-Echo testing

If the pile is linked to a structure above, then:

- Damping and other multiple reflected signals could mask the response from the tip.
- It is often difficult to mount the transducer on the side of the pile shaft.
- A clean, downwards-directed hammer blow is difficult to achieve.

When combined with the uncertainty of the value for v_b, this results in a doubtful determination of pile length by this method. In order to overcome this uncertainty, a modification of the Sonic-Echo method was developed, as described below.

One side of the pile is exposed over a length of at least 1.2 m below the pile cap. Two accelerometers are mounted on the sides of the pile, spaced at least 900 mm apart. The mounting system is an 'L-shaped bracket', bonded to the pile either by epoxy, bolting or magnetically. The pile cap is impacted by using a sledgehammer,

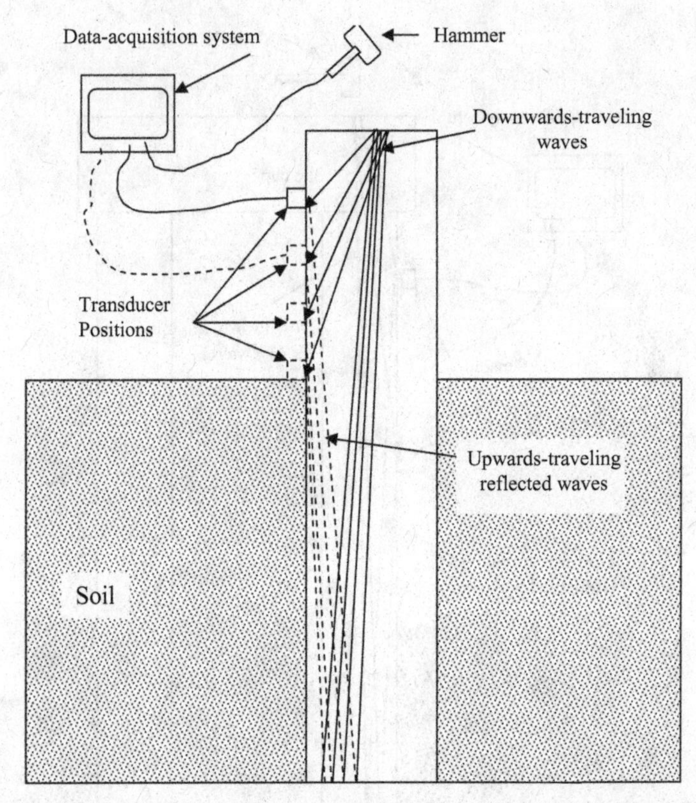

Figure 7.9 Schematic of the arrangement employed for Ultraseismic testing (after Olson *et al.*, 2001)

sending a stress wave down the pile. The upper accelerometer acts as a trigger for the recording system and both accelerometers record pile velocity over time. This method has only become possible in the late 1990s, thanks to the development of twin-channel data acquisition systems with sampling rates as high as 100 kHz.

The response signal from the pile tip is amplified as necessary to obtain a clear response in the accelerometer records. The time-difference between the stress-wave arrivals at the two accelerometers (both for the initial and return waves) will allow a calculation of the bar-wave velocity of the stress wave down the pile. This velocity will be used as v_b in the calculations for pile length. The relative arrival times of the signals at the two transducers can also allow the operator to discriminate between downwards- traveling impulses from the hammer impact, upwards-traveling waves that have been reflected and downwards-traveling waves that had originally traveled up into the superstructure and been reflected back down from there. The test is repeated at least ten times to obtain reproducible results. In this manner, the measured pile length can be predicted to within ± 2 %. It must be emphasized that the 'test pits' around the pile heads must be deep enough to expose at least 1.2 m of one side of the pile to be tested. Figure 7.8 presents a schematic of the arrangement employed for dual-sensor Impulse–Echo testing.

One of the few firms that has managed to obtain government funds for research into nondestructive testing of deep foundations is Olson Instruments, of Wheat Ridge, Colorado, USA. Olson Instruments personnel have also experimented with using two or more receiving transducers in a variation of the Impulse-Echo method which was developed as part of an assessment of test methods suitable for evaluation of unknown bridge foundations, sponsored by the FHwA from 1996 to 2001 (Olson, 2001). Olson's variant, dubbed the 'Ultraseismic Method', uses multiple locations for a pair of tri-axial transducers to create numerous intersecting ray paths that can be analyzed by established geophysical techniques to assess the location and significance of the reflectors. Using a similar analysis procedure to that described above, the Ultraseismic technique shows some promise for the assessment of unknown foundations beneath existing structures (Figure 7.9).

7.1.6 SAMPLE SPECIFICATION

For the reader who is considering the specification of nondestructive integrity tests for deep foundations, a sample specification for the Impulse-Echo method is given in Appendix IV of this book.

8

Sonic Mobility (Impulse Response)

As with the Sonic-Echo test described in the previous chapter, the Sonic-Mobility test first appeared in the literature in 1968 as 'La Méthode d'Admittance Mécanique', translated as the 'Mechanical Mobility method' (Paquet, 1968). Paquet's paper dealt with the application of electric transmission line theory to the one-dimensional problem of compression stress bar-waves transmitted from the pile head down its shaft. He explored the limitations of the method, particularly with respect to excitation frequency bandwidth. Bar-wave analysis in both time- and frequency-domains was discussed.

The test equipment used at that time was limited to analog technology, with pile-head excitation by swept-frequency vibrators mounted on the head of the foundation shaft. The 'Vibration test', as it was called then, was modified in the 1970s with the arrival of on-site computers and dedicated software, which allowed the use of impact devices in place of the cumbersome vibrator. Since that time, the Sonic-Mobility method has received several names, leading to some confusion, and has variously been referred to as shock, transient dynamic response (TDR), Impulse Response, Impulse Response Spectrum (IRS) and various others.

At that time, the predominant concrete foundation shaft construction method in France was the bored cast-*in situ* pile (known variously in different parts of the world as caisson, drilled shaft or cast-in-place pile). Paquet and Briard (1976) showed that for bored piles or drilled shafts the best approach was to analyze the vibration response in the frequency-domain, particularly if defects in the top few meters were to be detected. At this point, a divergence occurred between the frequency-domain approach developed in France and applied in the UK predominantly to drilled shaft construction

Nondestructive Testing of Deep Foundations B.H. Hertlein and A.G. Davis

Figure 8.1 Instrumentation set-up for the Vibration test on a drilled shaft

(Davis and Robertson, 1975, 1976) and the time-domain approach adopted in Holland and the Scandinavian countries where pre-cast driven piles were the norm and time-domain analysis was adequate for piles with constant cross-section (Reiding *et al.*, 1984; van Weele *et al.*, 1987). A factor that sometimes influenced the selection of one method over the other was that the vibration test required quite elaborate preparation of the shaft head to provide a relatively large smooth, level surface for the attachment of the vibrator and the transducers (Figure 8.1), whereas the Sonic-Echo test required comparatively little preparation, other than the removal of any laitance or loose concrete.

In either case, the control systems and data acquisition equipment were analog and very bulky compared with modern electronic equivalents. The vibration test equipment occupied a full 19-in rack, and was usually truck-mounted, due to its size (Figure 8.2). The equipment was certainly not considered portable!

Both theoretical research and the practical application of vibration testing to piling continued through the 1970s at the French National Institute for Building and Civil Engineering (Centre Expérimentale de Récherche et d'Études du Batîment et des Travaux Publics, acronym CEBTP) in Paris under funding from the French Federation of the Construction Industry (Briard, 1970; Davis and Dunn 1974; Davis and Robertson, 1976). The word was slowly getting out that in certain conditions, this nondestructive method could be used both as a quality control tool for new foundation construction and for forensic investigation of deep foundation shafts with problems.

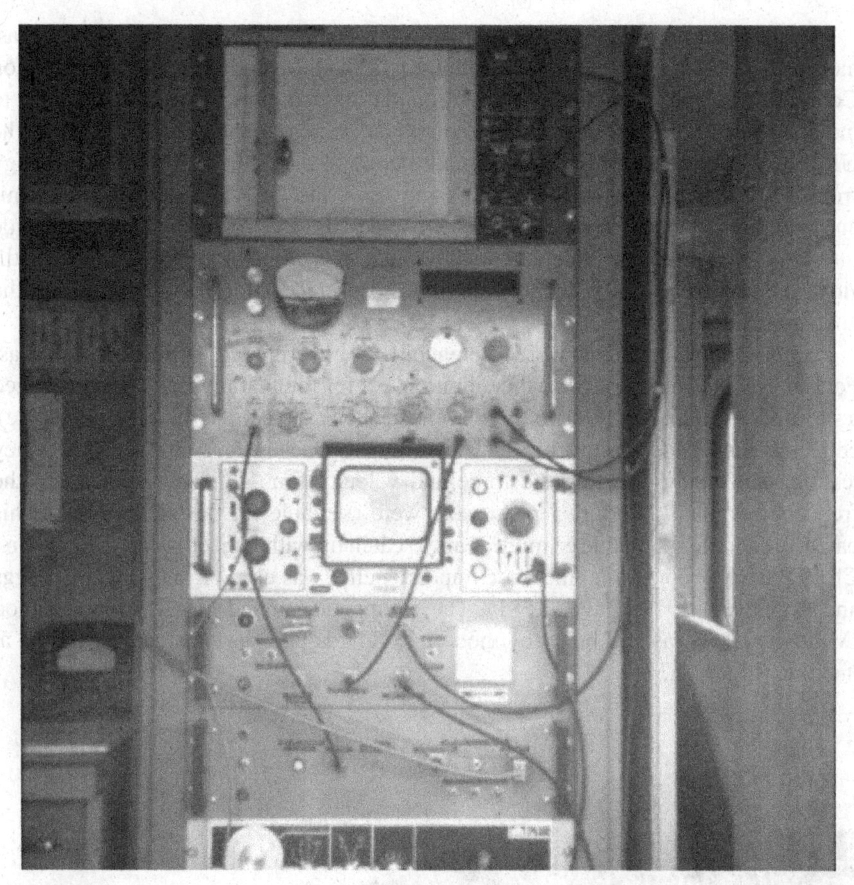

Figure 8.2 The control and data-acquisition equipment for the Vibration test – circa 1970

A number of significant papers were published on the topic between 1973 and 1976 but they were largely discussions of theory or applied research in France (Gardner and Moses, 1973; Davis and Dunn, 1974; Davis and Guillermain, 1974; Davis and Robertson, 1975, 1976; Robertson, 1976; Paquet and Briard, 1976; Guillermain, 1979). The first serious test of the validity and the limitations of the method outside France came with the testing of piles beneath two large concrete rafts supporting oil storage tanks in Southern England (Civil Engineer, 1974; Anon, 1975). The publicity generated by this case convinced the Construction Industry Research and Information Association (CIRIA) in the UK in 1977 that there was a need to prepare a document describing the existing nondestructive test methods for deep foundations (Weltman, 1977). At the same time, the French National Code on foundation practice included the provision that the allowable concrete stress in drilled shafts could be raised by 20 % if at least 25 % of the shaft population were tested nondestructively immediately after construction (DTU 1978).

Both the Vibration and Sonic-Echo methods became more widely known in the last half of the 1970s, even to the extent that public bodies such as the Greater London Council Boroughs in England required foundation contracts above a certain size to include quality control testing by these methods. A large foundation contractor in the UK ran its own nondestructive testing auto-control teams, licensing the technology from TNO in Holland. At about the same time, the CEBTP established a semi-autonomous commercial division (Testconsult-CEBTP Ltd) in England to provide integrity testing for deep foundations as an independent laboratory. Testing was still slow and relatively costly, because of shaft-head preparation requirements and the analog technique.

Paquet applied for and obtained a patent in 1974 covering the application of the Fast Fourier Transform algorithm to the vibration test for foundation shafts. He envisaged a short-duration impact force to the shaft head such as a hammer blow (white noise), collecting the response to that impact with a velocity transducer, as in swept-frequency testing, and then converting the force and velocity time-domain responses to the frequency-domain. No digital processors were available for this at the time of his patent. Technology took three more years to catch up with Paquet's lead, and the first CEBTP 'shock' apparatus (méthode impulsionelle) was used on site in 1977 (Higgs and Robertson, 1979; Davis, 1981). The equipment, named MIMP from 'Method IMPulsionelle', was still bulky by modern standards but was considered portable at the time (Figure 8.3).

Figure 8.3 The 'MIMP' equipment set up to test a drilled shaft

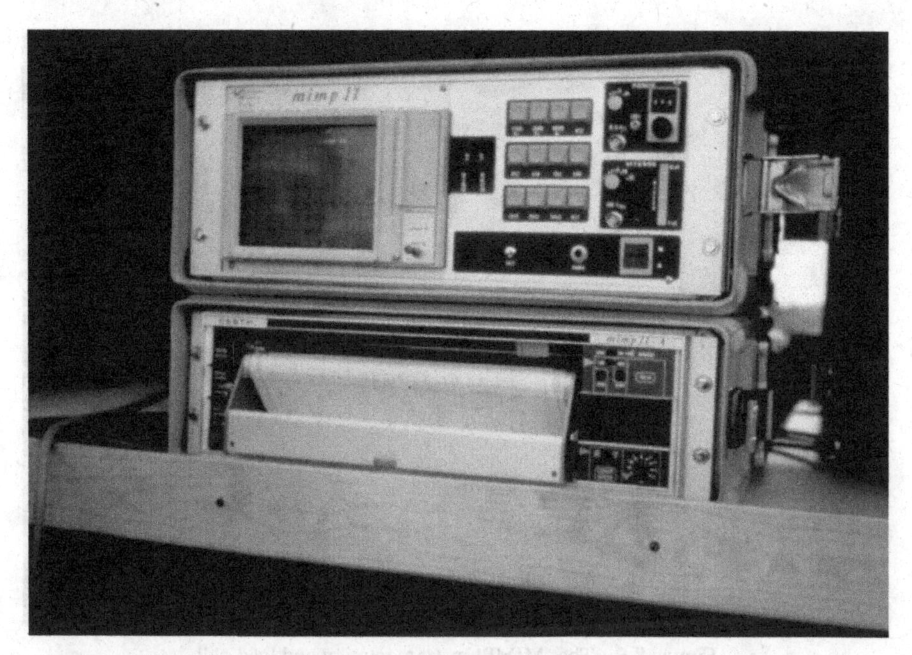

Figure 8.4 The 'MIMP' acquisition unit (upper) and printer (lower)

All the advantages of frequency analysis were maintained and shaft-head preparation was reduced to ensuring that the concrete in the shaft head was sound and uncracked. The computer processor, a Digital PDP-11, was housed in a purpose-built unit that included a hard-copy printer but had no data storage (Figure 8.4). A load-cell was placed on the pile head and struck with a 1-kg (2-lb) hammer (Figure 8.5). The load-cell triggered the data acquisition cycle, as it still does today, but now the load-cell is built into the hammerhead.

This equipment made testing much faster and more economical than with the swept-frequency vibration method and so the Impulse-Response method was born (Swann, 1983; Stain and Davis 1983). The commercial advantages of the new method were readily apparent and competitors soon broke the patent, with no reaction from CEBTP. The development of scientific instruments for general vibration analysis put spectrum analyzers in the hands of would-be testers and by 1985 (Bracewell, 1986), several testing companies operated an Impulse-Response system in one form or another (Ellway, 1987; Olson and Wright, 1989). In the English-speaking countries, the method was referred to as the Transient Dynamic Response (or TDR) method for several years, until conflict of the name with 'Time-Domain Reflectometry', used in electrical engineering, required a change. Various names, such as Impedance or Sonic Mobility were used. Finally, a general consensus in the USA around 1990 decided on the name Impulse Response.

Control of data input and limited data storage for fieldwork became possible around 1982 with the use of small digital computers such as the Hewlett Packard HP-85 linked

Figure 8.5 The 'MIMP' instrumentation and load-cell

to the data-acquisition system or spectrum analyzer on site. In addition, at about the same time, PCB Piezotronics in the United States introduced their impulse hammer to the European foundation testing community. The impulse hammer contained a built-in force transducer to replace the cumbersome load-cell and separate hammer that were then the state-of-the-art (Figure 8.6). Greater strides were made in 1985 with the arrival of relatively portable personal computers (PCs), capable of mass data storage, and analog-to-digital (A/D) data acquisition cards with very high sampling rates and on-board pre-trigger facilities (Figure 8.6). An important advantage of these new systems was that they allowed storage of all test data and subsequent analysis in the relative calm of the office after testing, with increased confidence in the final result (Figure 8.7). Even more significantly, this technological advance had a very great influence on the extension of foundation test simulation methods, and also facilitated the application of the Impulse-Response method to other fields of structural investigation besides foundation testing (Chan et al., 1987; Davis and Kannedy, 1998; Davis, 2003).

Paquet introduced foundation shaft simulation methods in 1970 but such matching techniques required the digitization of actual foundation test responses for comparison with simulated responses (Davis and Dunn, 1974). Only with the arrival of easy digital storage of the real test response could simulation methods be fully exploited (Davis, 1997). New possibilities for modeling as a result of full data storage led to the development of the Impedance-Log profiling method (Olson and Wright, 1989; McCavitt and Forde, 1990; Davis and Hertlein, 1991).

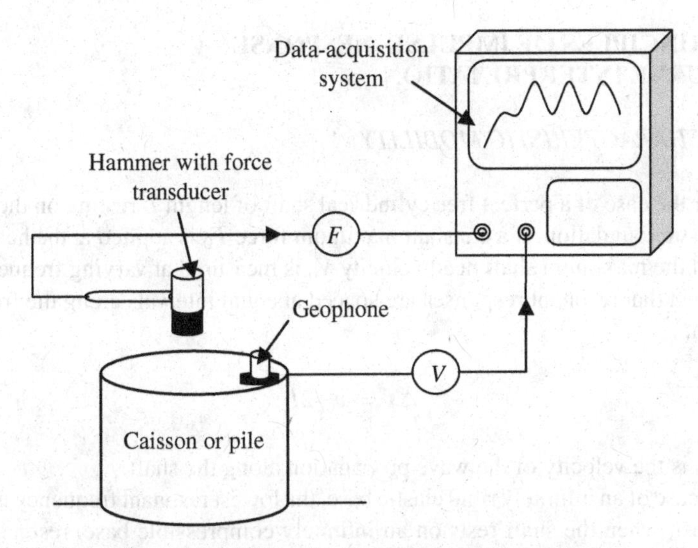

Figure 8.6 Schematic of an Impulse-Response test

In order to appreciate the possibilities and limitations of this method when applied to deep foundation testing, it is necessary to understand the theory behind the Impulse-Response method. The basic theory was advanced by Paquet in his paper 'Vibration Study of Concrete Piles: Harmonic Response' (Paquet, 1968), and has changed little since that time (see Appendix IV).

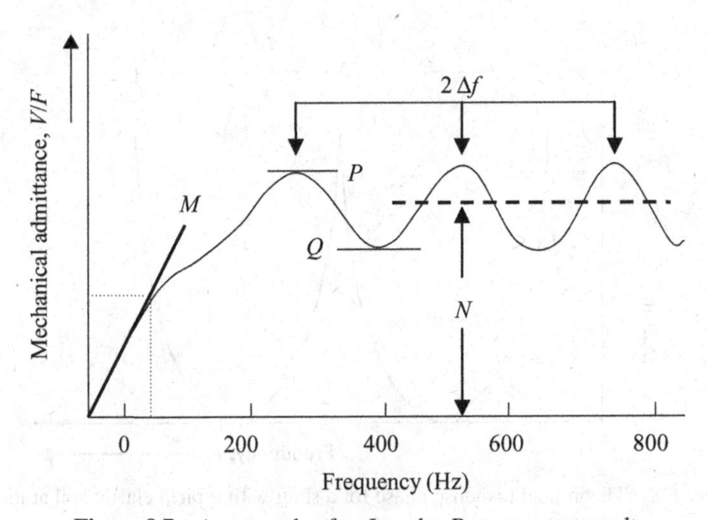

Figure 8.7 An example of an Impulse-Response test result

8.1 PRINCIPLES OF IMPULSE–RESPONSE CURVE INTERPRETATION

8.1.1 CHARACTERISTIC MOBILITY

Consider the case of a perfect free cylindrical shaft of length L resting on the surface of an elastic foundation. If a constant maximum force F_0 is applied at the head of the shaft and the maximum shaft head velocity V_0 is measured at varying frequencies, it is observed that resonant responses are spaced at equal intervals along the frequency spectrum:

$$\Delta f = v_c/2L \tag{8.1}$$

where v_c is the velocity of the wave propagation along the shaft.

In the case of an infinitely rigid elastic base, the lowest resonant frequency is $v_c/4L$. In contrast, when the shaft rests on an infinitely compressible base, resonance first occurs at a very low frequency, approaching zero. When the base is an elastic soil of normal compressibility, the lowest resonant frequency lies at an intermediate position between the infinitely rigid and the infinitely compressible, as shown in Figure 8.8.

When the shaft is embedded in soil, the movement of the shaft is damped by the presence of the soil. The response curve $|V_0/F_0|$ is attenuated and the shape of this curve is as shown in Figure 8.7. The denser the soil and the longer the shaft, the greater is the attenuation, so that the differential between the maxima and the minima on the response curve is reduced.

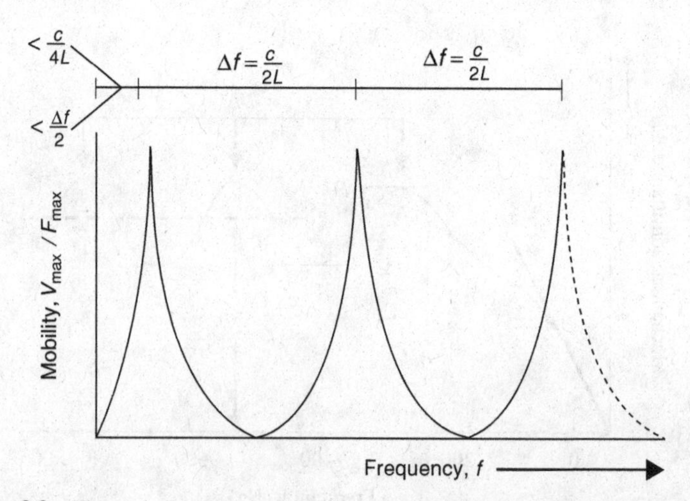

Figure 8.8 Theoretical response phase for a shaft with typical elastic soil at its base

A factor that gives a measure of the soil damping effect was given by Briard (1970) as:

$$\sigma = (\rho'\beta l)/(\rho_c v_c r) \tag{8.2}$$

The mean value of $|V_0/F_0|$ in the steady-state region of the response curve is known as the 'mechanical admittance', or as the 'average mobility', N, and theoretically:

$$N = 1/(\rho_c v_c A_c) \tag{8.3}$$

This is the inverse of 'mechanical impedance'. The maximum value, P, and the minimum value, Q, of $|V_0/F_0|$ provide a measure of the soil damping effect where:

$$P = N\coth(\sigma L) \tag{8.4}$$

$$Q = N\tanh(\sigma L) \tag{8.5}$$

N can therefore be expressed in terms of P and Q as:

$$N = \sqrt{(PQ)} \tag{8.6}$$

and σL may be calculated from:

$$\coth(\sigma L) = \sqrt{(P/Q)} \tag{8.7}$$

If the exact length of a shaft is known, it is possible to determine v_c by measuring Δf and substituting into equation (8.1). The propagation velocity is related to the elastic modulus and the density of the concrete and so a determination of v_c gives a measure of concrete quality. Conversely, if the average concrete quality is known, for example, by determination of velocity measurements on representative concrete cores, it becomes possible to determine the effective length of a shaft by measuring Δf. This can then be checked against specific length.

If the shaft is discontinuous or severely broken, then only the length down to the discontinuity will be measured. This applies to the presence of significant bulbs (bulges) or neck-ins on bored shafts and the total length of the shaft is not measured.

Owing to the cumulative effect of soil damping on a long shaft, a test on a shaft with a length/diameter ratio (L/d) greater than 30 is unlikely to be very definitive, unless the shaft passes through a very soft soil onto a rigid stratum.

The area of cross-section of the shaft is calculated from equation (8.3) provided that ρ_c and v_c are known. If the length determined from Δf is nearly equal to that specified, then the value of v_c will be known. The measured value of N is predominantly a function of the properties of the upper portion of the shaft. If N is much greater than its calculated value, it is likely that the upper part of the shaft is defective, either because of restricted cross-sectional area or inferior concrete.

If equations (8.1) and (8.3) are combined, an expression for the mass of the shaft is obtained that includes only parameters actually determined from the response curve:

$$M_p = LA_c\rho_c = 1/(2\Delta f N) \tag{8.8}$$

8.1.2 MEASUREMENT OF SHAFT STIFFNESS

When the shaft is excited at low frequencies, the inertia effects are insignificant and the shaft–soil unit, behaving like a spring, gives a straight-line response at the start of the $|V_0/F_0|$ curve. The inverse of the slope of this straight line measures the apparent dynamic stiffness of the head of the shaft. The dynamic stiffness is given by:

$$E' = (2\pi f_M)/|V_0/F_0|_M \tag{8.9}$$

where f_M and $|V_0/F_0|_M$ are the coordinates of the point 'M' on the response curve and E' corresponds to the slope of the initial tangent modulus to a load–displacement graph obtained from a static load test on a shaft. This parameter can be used to compare the load carrying ability of a population of shafts on any one site.

The apparent stiffness of the shaft head is a function of:

- The stiffness of the concrete in the shaft, E_c.
- The stiffness of the soil surrounding the shaft as indicated by the attenuation of the oscillations of the response curve and measured by the damping factor.
- The stiffness of the soil supporting the base of the shaft.

By knowing the first two properties, it is possible, in principle, to determine the apparent stiffness measured at the shaft head in the following two extreme cases.

A shaft supported on an infinitely rigid base would give the maximum stiffness possible and may be calculated from:

$$E'_{max} = E'_{\infty} \coth(\sigma L) = E'_{\infty}\sqrt{(P/Q)} \tag{8.10}$$

with:

$$E'_{\infty} = (A_c E_c \sigma L)/L \tag{8.11}$$

A shaft with no base support, behaving as if cut off at depth, would give the minimum possible stiffness and may be calculated from:

$$E'_{min} = E'_{\infty} \tanh(\sigma L) = E'_{\infty}\sqrt{(P/Q)} \tag{8.12}$$

These two values provide upper and lower bounds of stiffness with which the actual measured stiffness, E', may be compared.

8.2 PRACTICAL CONSIDERATIONS

8.2.1 ACOUSTIC LENGTH

The resonating length, L, and therefore the depth to the resonating feature in the shaft, can be calculated from the field test mobility plot using equation (8.1) above, by measuring the resonant frequency interval Δf. From equation (8.1), it is evident that

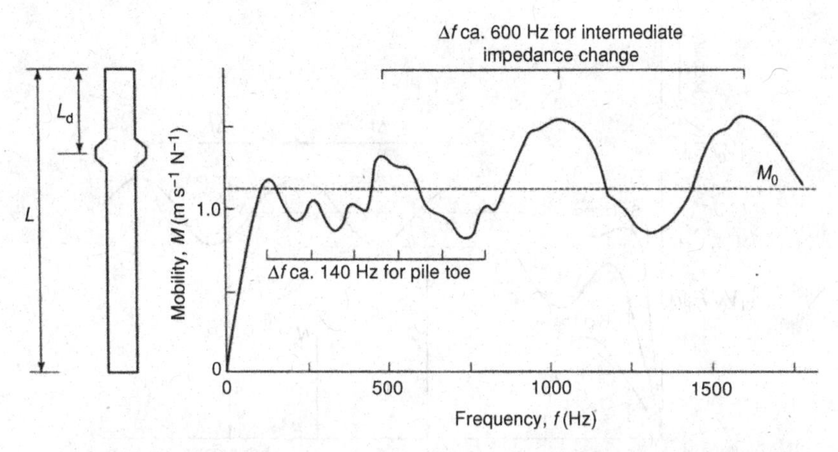

Figure 8.9 Intermediate response superimposed on response from shaft base

the smaller the resonant frequency interval, then the greater the depth to that feature causing the resonance. Thus, a feature above the shaft toe will produce a greater Δf value than that for the shaft length. If this feature produces near-total reflection of the downwards traveling bar wave, it masks completely any feature below it, including the shaft toe. However, if the first-encountered feature only partially reflects the downwards traveling wave (as often for partial loss of shaft section), the remainder of the wave continues down the shaft to the shaft toe and is reflected back to the shaft head. The two wave fronts then tend to interfere, and overlay one another on the mobility–frequency trace (see Figure 8.9). When this occurs, experience is needed in interpretation of the resultant trace.

8.2.2 *FREQUENCY SHIFT OF MOBILITY PLOT*

Davis and Dunn (1974) demonstrated that the nature of any pile impedance change, whether it is the pile toe or an intermediate feature such as a loss of section or a bulb, can be deduced from the position of the resonating peaks relative to the origin. If the first peak is at a distance Δf, or a multiple of Δf, from the origin, then this indicates a free-end reflection type, such as for a complete section loss (e.g. crack across entire cross-section). On the other hand, if the resonant peaks are located at $\Delta f/2$ from the origin, this would indicate a fixed-end response such as a shaft enlargement or a pile socket into very stiff material (see Figure 8.10).

8.2.3 *PILE STATIC/DYNAMIC STIFFNESS RELATIONSHIP*

Davis and Dunn (1974) suggested that there is a relationship between the low-frequency dynamic stiffness measured in the Impulse-Response test and the static

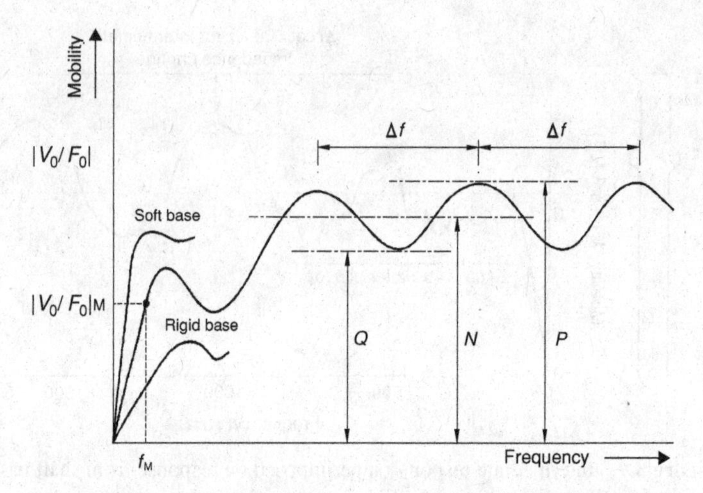

Figure 8.10 Impulse-Response curve – mobility against frequency

stiffness deduced from the initial elastic portion of the load–settlement curve for a deep foundation load test when 100 % recovery is obtained. They postulated that the dynamic stiffness would be higher than the static stiffness by a factor, k. According to Davis and Dunn, k is dependent on the site soil conditions, foundation shaft type and size (L/d ratio) and is thought to vary between 1.5 and 2. The reason for the higher dynamic stiffness is considered to be the smaller strains generated in the dynamic test, together with a small contribution from differing strain rates in the two tests. In addition to the data given in the original paper (Davis and Dunn, 1974), data from other site studies have given support to this theory (Davis and Robertson, 1976; Davis, 1985).

Although this point has not been universally accepted, such a concept is typically used in site-testing programs to assess shafts of similar dimensions and separate those shafts with uncharacteristically low dynamic stiffness for further investigation. It must be stressed that the measured low-strain dynamic stiffness for a foundation shaft has no relationship to that pile's load-carrying capacity.

McCavitt and Forde (1990) proposed that the 'effective mass' of the shaft (how much of the shaft/soil system is being excited by the Impulse-Response test) is a more useful parameter. They obtained the effective mass, m' from a plot of the shaft's 'inertance' against frequency. Figures 8.11 and 8.12 show the relationships obtained between dynamic stiffness and effective mass, respectively, against the L_d/L ratio for shafts at the Blyth, Northumberland, UK, Class-A test site, where L is the shaft length and L_d is the length to any detected defect in the shaft. They suggested that the effective mass parameter can help in distinguishing between reflections from either loss or gain in shaft section. Few additional data are available at present to support this postulation.

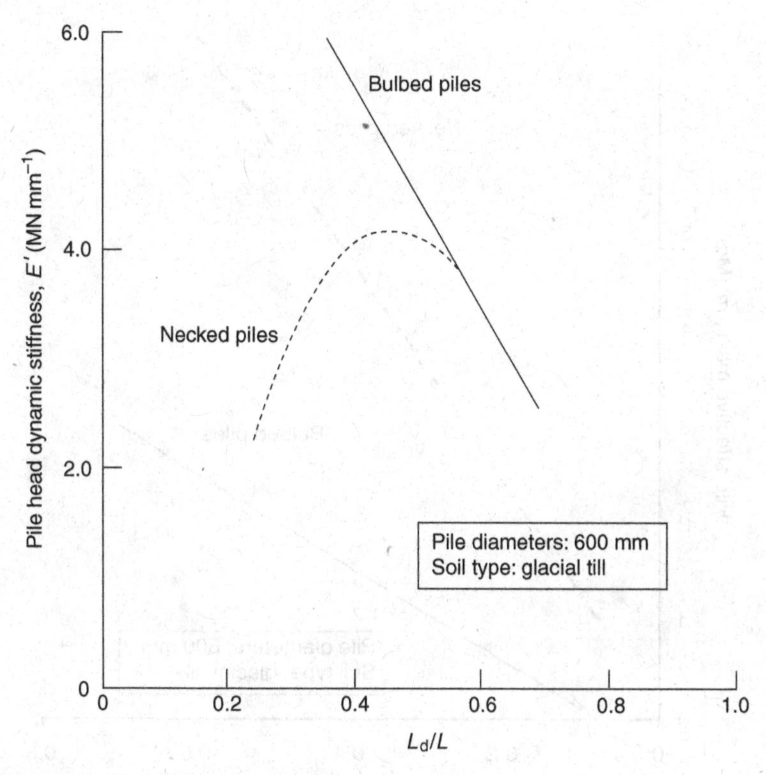

Figure 8.11 Relationship between dynamic stiffness and L_d/L ratio for shafts with neck-ins and bulbs

8.3 CLASSIFICATION OF SIGNAL RESPONSES

Detailed interpretation of complex signal responses from Impulse-Response tests has always been the domain of trained and experienced specialists. It is not likely that this situation will change in the future. However, when these tests are used for routine quality control testing, it is desirable that most of the signals are understandable to other non-specialist engineers associated with the construction project. The different parties to any deep foundation construction contract must agree to the validity of the basic information given by the tests if the testing program is to be useful.

Care should be taken to ensure that the results of the integrity testing do not serve as the sole basis for acceptance or rejection of any particular shaft. Rather, other indicators, such as soil profiles, construction records and observations, should be considered as well. However, if an engineer perceives the interpretation of the Impulse-Response or Sonic-Echo test data to be subjective, he or she will have little confidence in their use as quality control tools. To attempt to resolve this problem, the Construction Industry Research and Information Association (CIRIA) in the UK

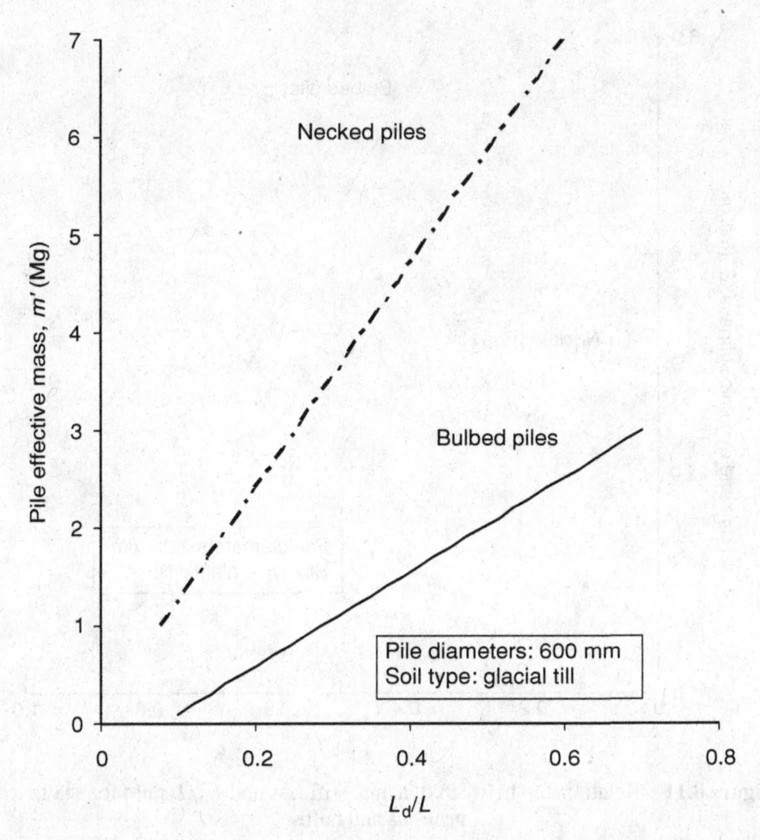

Figure 8.12 Relationship between effective mass and L_d/L ratio for shafts with neck-ins and bulbs

proposed a signal classification system for both Sonic-Echo and Impulse-Response test results, to differentiate the simpler signal responses from those that are more complicated (Turner, 1997). The discussion that follows applies equally to Sonic-Echo and Impulse-Response testing.

The CIRIA proposes that the signal response of a foundation shaft can be classified into one of three main categories, based upon an evaluation of the number of significant impedance changes identifiable within the shaft that cause portions of the input signal to be reflected back to the shaft head. These three signal response types are placed in Categories 0, 1 and 2, as described in the following.

8.3.1 TYPE 0 SIGNAL

A type 0 signal is one in which the damping effect of the surrounding soil attenuates any return signals to such an extent that the toe cannot be discerned. Therefore,

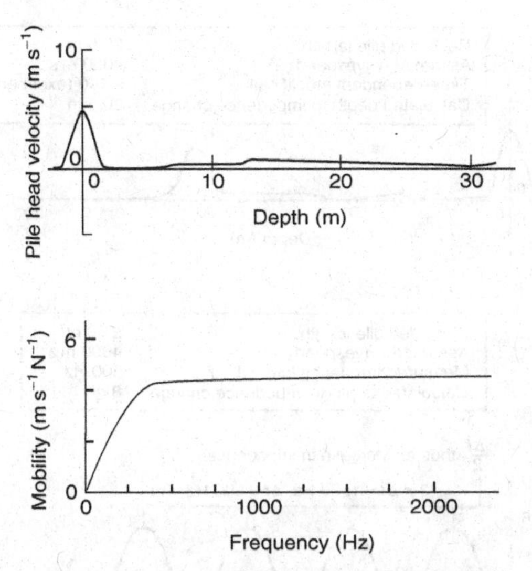

Figure 8.13 Examples of type 0 Impulse-Response signals

there is no significant impedance change within the shaft capable of detection within the effective penetration depth of the system. The non-specialist engineer would easily appreciate the reason for this, provided that the basic principle of the test was understood.

Figure 8.13 illustrates typical examples of type 0 signals for time-based and frequency-based systems.

8.3.2 TYPE 1 SIGNAL

A type 1 signal contains one clear, major response, indicating that the shaft is responding to a single acoustic input. This is for a shaft containing a single major impedance change, either the shaft toe or some significant intervening feature. No other significant response would be visible on the recorded trace. Type 1 signal responses should be very similar to the theoretical simple signal expected from the test and be easily recognizable as well as easily simulated by computer.

Typical type 1 signal responses would be like those shown in Figure 8.14.

8.3.3 TYPE 2 SIGNAL

A type 2 signal is one containing more than one major response, so that the interaction of overlapping responses from different levels within the shaft makes interpretation of the resulting response a complex matter. At one extreme, type 2 signals might display

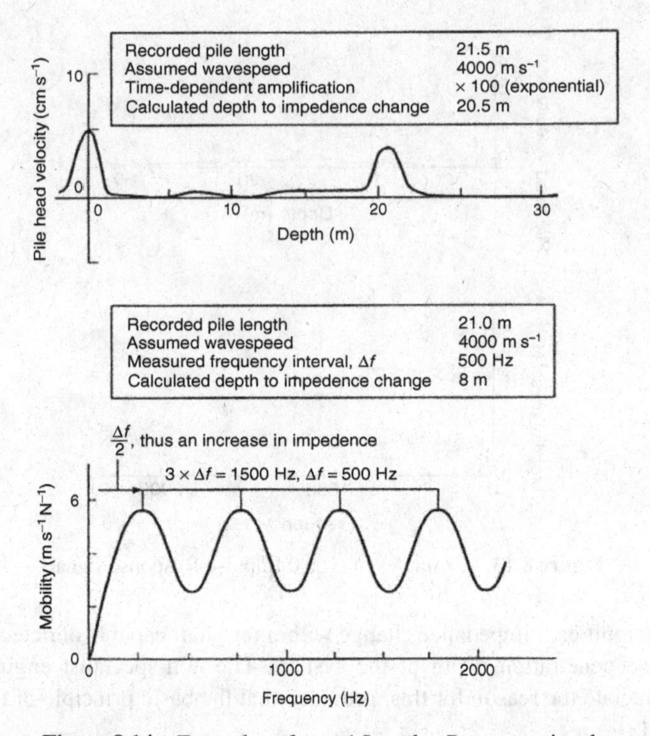

Figure 8.14 Examples of type 1 Impulse-Response signals

a clear major response from the length of the shaft responding as a major acoustic unit, but with intermediate responses to local changes in shaft impedance within that acoustic unit, as shown in Figure 8.15. At the other extreme, type 2 signals might contain no clear major response to indicate if part of the shaft is responding as a single acoustic unit, as indicated in Figure 8.16.

A type 2 signal would require interpretation by a specialist, because simple models do not easily explain the response.

The CIRIA suggest that a classification in these terms of the signal responses obtained on an individual site will help to assess how much weight should be given to the test results from the shafts on that site. The quality of the test results will depend upon:

- The characteristics of the test system, particularly its dynamic range, its resolution and its signal-to-noise ratio.
- The shaft characteristics, especially the length/diameter ratio, the quality of the shaft material and the shape of the shaft.
- The nature of the surrounding soil: the stiffer the soil, the greater the signal attenuation. In addition, a boundary between soils of different relative stiffness acts as a reflective layer or impedance change within the shaft–soil system.

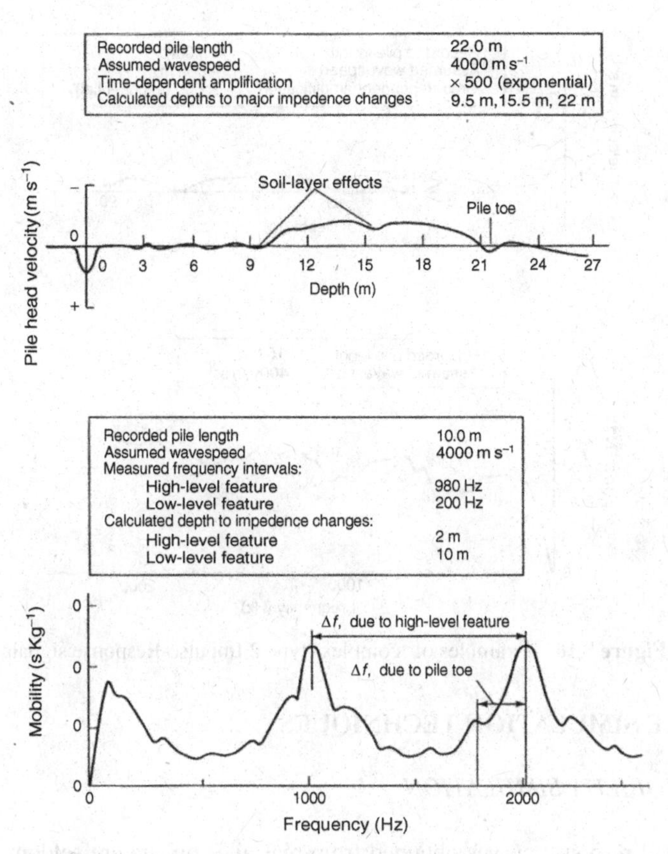

Figure 8.15 Examples of 'clear' type 2 Impulse-Response signals

All nondestructive testing is, by definition, an indirect way of examining a structure's material properties. In order to prove the validity of the interpretation of these methods, they have to be put to the test in very carefully controlled exercises. These are referred to as 'Class-A' studies in the geotechnical industry, and several such studies have put NDT methods to the test. Chapter 11 reviews these exercises and their effect on the NDT industry.

The Impulse-Response method has been tested several times in this way. Most of these exercises concluded that test methods relying on stress-wave generation at the shaft head (Sonic Echo and Impulse Response) were limited when the pile length/diameter ratio became large, particularly in stiff soils. The FHwA study reported by Baker *et al.* (1993) concluded that for piles and shafts with critical loads transmitted through their bases, these methods should not be used for quality control of new construction, because of the lack of certainty of base integrity information received. As a result, these methods are not in common use for quality control in North America, but still see extensive application in Europe and Asia.

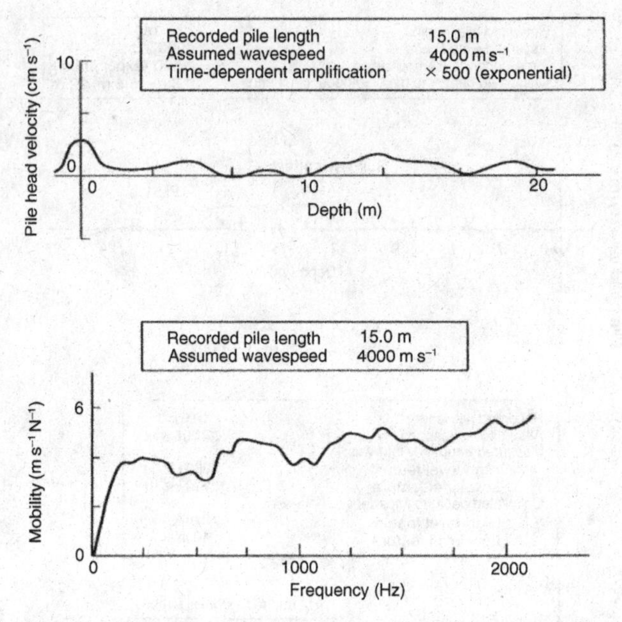

Figure 8.16 Examples of 'complex' type 2 Impulse-Response signals

8.4 PILE SIMULATION TECHNIQUES

8.4.1 MOBILITY SIMULATION

The $|V_0/F_0|$ response curves obtained from real piles on site are seldom as simple as the theoretical curve for the perfect pile. Numerous factors influence the response curve, among which the most common are:

(a) Variations in the pile shaft diameter.
(b) Variations in the pile concrete quality.
(c) Variations in the stiffness of the soil surrounding the shaft.
(d) Exposure of the pile head above ground.

The general effect of these factors is to produce an oscillation that is superimposed on the normal oscillation of the $|V_0/F_0|$ response curve (Figure 8.16). This makes interpretation more difficult. In order to appreciate the effect of such variables on response and to improve interpretation, the response curve of any pile with a known anomaly in the shaft may be predicted by simulating mathematically the pile response. In the early days (circa 1970), use was made of the analog existing between mechanical wave propagation and electric transmission line theory, the mathematics of which were well understood at that time (Davis and Dunn, 1974).

Consider a perfect cylindrical unit-length segment of a pile. The force F_t applied to the top of the segment may be represented by the current I_t and the velocity at the

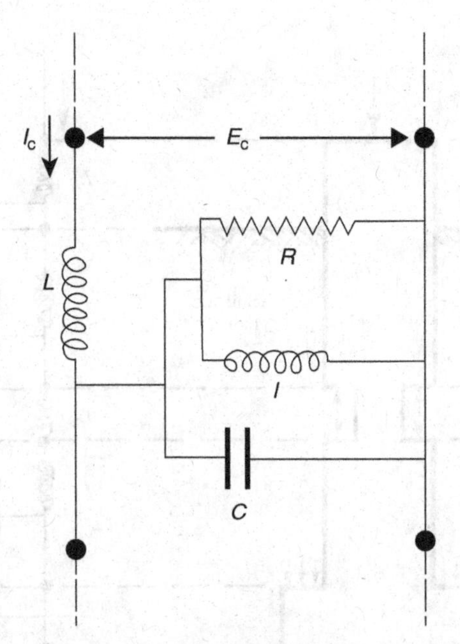

Figure 8.17 Electrical analog of a section of pile (after Davis and Dunn, 1974, reproduced with permission)

top of the segment by the voltage E_t. This means that the mechanical impedance of the pile segment corresponds to the electrical impedance I_t/E_t. The other electrical analogs may be summarized as follows:

- The mass of the pile per unit length is represented by the capacitance, $C = \pi r^2 \rho_c$.
- The soil resistance per unit length is represented by $1/l = \pi\rho'\beta'^2$, where l is the inductance of a coil.
- The energy lost by the soil-damping effect is represented by the admittance of resistance, $R(1/R = 2\pi r\rho'\beta')$.
- The stiffness per unit length of pile is represented by an inductance, L, where $1/L = \pi r^2 E_c$.

As an example of how a defective pile may be represented by an electrical analog, Figure 8.17 shows the electrical elements that could be used as a first approximation to represent each unit length of the different pile sections.

Figure 8.18 shows the compound analog of all segments in the pile. The base stiffness is represented by an inductance, L', where the base stiffness is given by:

$$1/L' = 1.84r[E_b/(1 - \nu^2)] \tag{8.13}$$

By using a computer to solve the matrix that takes the general form:

$$|V/F| = |a|b|c|d|e|f||V_B/F_B| \tag{8.14}$$

where $|V_B/F_B|$ is the soil admittance at the pile base, it is possible to calculate the form of the response curve for any pile.

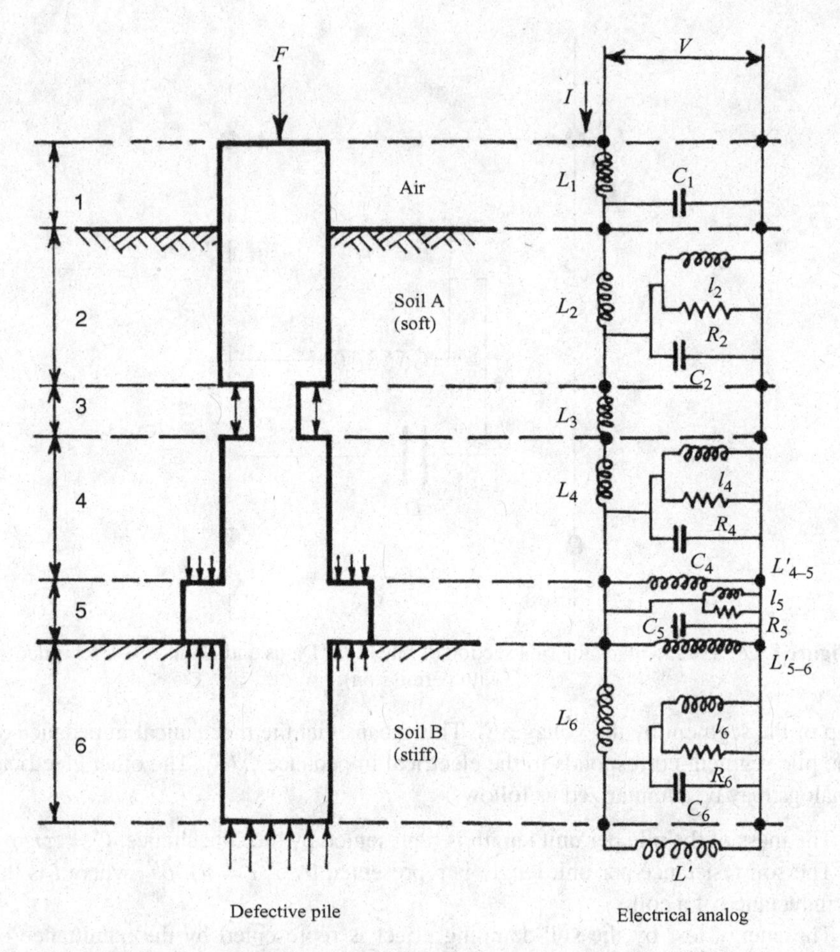

Figure 8.18 Compound electrical analog of all segments in a pile (after Davis and Dunn, 1974, reproduced with permission)

Figure 8.19 shows a computer plot of a theoretical response curve for a hypothetical pile with varying cross-sections. The distance between the primary maxima enables the length of the enlarged pile head to be estimated while the distance between the secondary maxima enables the total pile length to be estimated. The enlarged pile head diameter may be determined approximately from equations (8.4) and (8.5) by taking Q as the minimum and P as the maximum values of $|V_0/F_0|$. However, the reduced diameter of 600 mm can only be estimated approximately by taking Q as a minimum, just before the maximum peak, and P as the maximum. No information can be learned about the pile-toe diameter from inspection of the response curve.

Figure 8.20 shows a response curve for a pile whose top portion consists of poor-quality concrete. It is possible to estimate the length of poor concrete and, given the total pile length, the approximate average value of v_c.

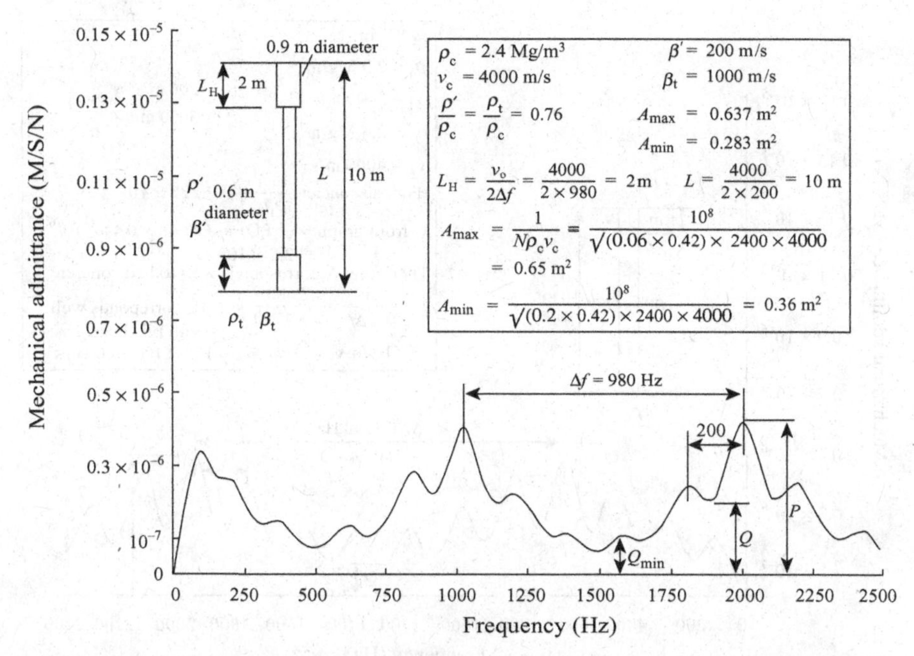

Figure 8.19 Computer simulation of a pile with varying cross-sections (after Davis and Dunn, 1974, reproduced with permission)

It is evident from these two examples that if a combination of defects or variable conditions exist in practice, it can be difficult to interpret the actual response curve without the facility of being able to study theoretically the effects of varying geometrical and material parameters on simulated response curves.

The analyst needs to have as much information as can be obtained about the real pile and soil conditions before attempting a simulation. This helps to reduce the number of steps required to obtain a matching fit between the real and simulated mobility–frequency response curves. Typical values of material properties that can be used as a first approximation are:

- Concrete bar-wave velocity from 3800 to 4100 m/s.
- Concrete density from 2300 to 2400 kg/m^3.
- Soil density from 1700 to 1900 kg/m^3.
- Shear wave velocity in soil around pile shaft from 100 to 300 m/s.

8.5 TIME DOMAIN–VELOCITY REFLECTORS

Filtering and smoothing of the velocity–time response signal can help in removing signal noise from the trace when carefully used. An alternative way to locate true velocity reflectors on the time trace is to perform a Fast Fourier Transform (FFT) on

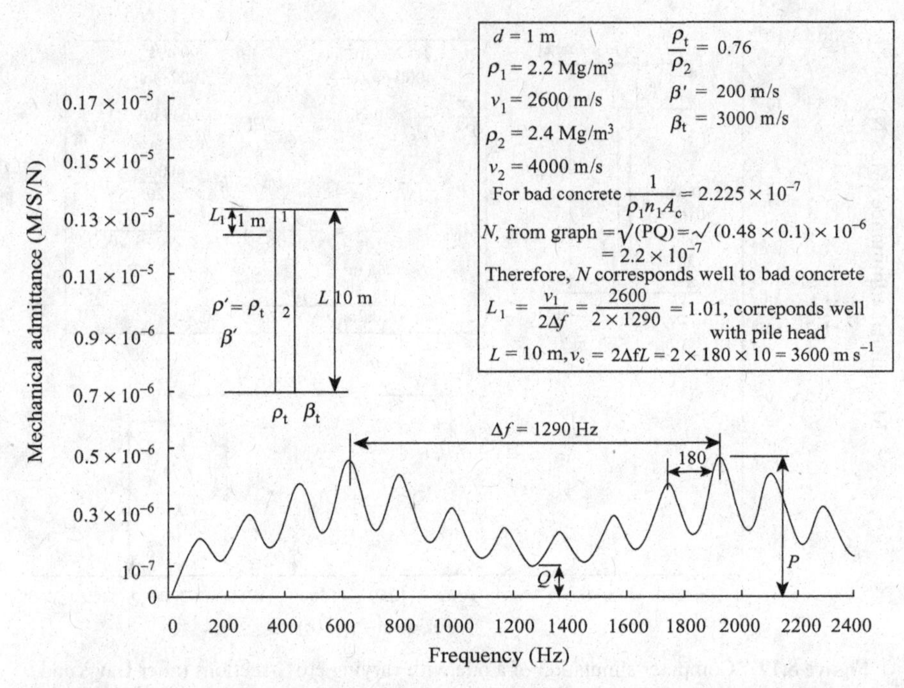

Figure 8.20 Example of a response curve obtained for a pile with poor-quality concrete at the top (after Davis and Dunn, 1974, reproduced with permission)

the time trace alone, and then to perform an Inverse Fast Fourier Transform (IFFT) on the FFT frequency signal array. This applies a smoothing effect on the resultant time trace, with the advantage of removal of some degree of background noise. However, this method does have certain limitations:

- The double treatment (FFT–IFFT) results in a time signal with a singular polarity. Identification of positive or negative impedance changes is lost and no distinction can be made between neck-ins and bulges.
- The influence of the Rayleigh wave at the pile head on the velocity response–time trace masks the response from approximately the upper 3 m of the pile shaft.

8.5.1 *SAMPLE SPECIFICATION*

For the reader who is considering the specification of nondestructive integrity tests for deep foundations, a sample specification for the Impulse-Response method is given in Appendix IV of this book.

9

The Impedance-Log Analysis

Drilled shaft testers have always had the dream of creating an image of the shaft as it actually is in the ground, 'warts and all'. The closest method to date for realizing this dream is the Impedance Log, developed by Paquet (1991, 1992), which can produce a probable shaft shape profile with depth by combining the amplified Sonic-Echo time-domain response with the characteristic foundation shaft impedance measured in the frequency-domain by the Impulse-Response method. However, because the mathematics in the Impedance-Log method are limited to a one-dimensional model, the resultant computed shape is symmetrical about the shaft axis and hence the true horizontal position of any asymmetrical anomaly around the pile circumference, such as a partial neck-in, cannot be reproduced. Other efforts to produce foundation shaft shapes from low-strain stress-wave data from either the Impulse-Response or the Sonic-Echo tests can be found in Mu and Zhao (1991), Honma et al. (1991) and Rix *et al.* (1993).

Paquet observed that even though the force applied to the shaft head in these test methods is transient, the wave generated by the blow is not. This wave picks up information about changes in shaft impedance as it proceeds downward and this information is carried back to the shaft head as the wave energy is partially reflected upwards by the changes in impedance. The Reflectogram so obtained in the Sonic-Echo test has been used to estimate the shape of piles (see Chapter 7). Paquet already pointed out in 1968 that use of the temporal signal from the sensor alone could not provide the shaft shape, because of the need to eliminate waves returned from changes in the surrounding soil profile (deconvolution); unless this is done, the Reflectogram so obtained cannot be quantified. However, it is possible to sample both wave reflection and the impedance properties of the tested shaft. Measurements of force and velocity response are stored as time-base data, with a very wide band-pass filter and rapid sampling. Resolution of both weak and strong response levels is thus favored. In

---

Nondestructive Testing of Deep Foundations B.H. Hertlein and A.G. Davis

the Reflectogram, a complete shaft defect (zero impedance) is equivalent to 100 % reflection, while an infinitely long shaft with no defects would give zero reflection.

If either a defect or the shaft tip is at a considerable distance from the shaft head, damping within the shaft reduces the reflected amplitude. With uniform lateral soil conditions, this damping function has the form $e - \sigma L$, where L is the shaft length and σ is the damping factor. The Reflectogram can be corrected by using such an amplification function to yield a strong response over the total shaft length, as is frequently carried out in the treatment of Sonic-Echo data. The stages involved in generating an impedance profile are shown in Figure 9.1.

The frequency-domain (impedance) analysis obtained from the Impulse-Response test confirms shaft length and gives the shaft dynamic stiffness and characteristic impedance, I, as follows:

$$I = \sigma_c A_c C_b \tag{9.1}$$

where σ_c is the density of the shaft concrete, A_c is the shaft cross-sectional area and C_b is the concrete bar-wave velocity.

In addition, simulation of the tested shaft and its surrounding soil can be carried out most efficiently in the frequency-domain. The Reflectogram and the characteristic impedance can then be combined to give dimensions to the Reflectogram and produce a trace referred to as the 'Impedance Log'.

Figure 9.2 shows a Relative Reflectogram calculated from the velocity–time reflections for a synthetic pile. The output of this analysis is in the form of a vertical section through the shaft, giving a calculated visual representation of the pile shape. The final result can be adjusted to eliminate varying soil reflections by use of the simulation technique.

Field-testing equipment must meet the following requirements:

- Hammer load-cell and the velocity transducer must have been correctly calibrated (within the six months prior to testing).
- Data acquisition and storage must be digital for future analysis.
- Both time- and frequency-domain test responses must be stored.
- Minimum sampling rate for most common pile shaft lengths should be 10 kHz, with a minimum of 1024 data points for each waveform.

Observing the signal response stages in Figure 9.1, the first analytical step is to remove the following from the velocity waveform:

- The motion of the top of the shaft caused by the hammer impact.
- The effects of the impedance change resulting from changes in the soil profile.

This is achieved by calculating a theoretical mobility plot for an infinitely long, defect-free shaft with a nominal shaft diameter and known lateral soil variations. This computed mobility plot is subtracted from the test response. This gives a 'reflected' mobility response containing information about changes in the shaft geometry and

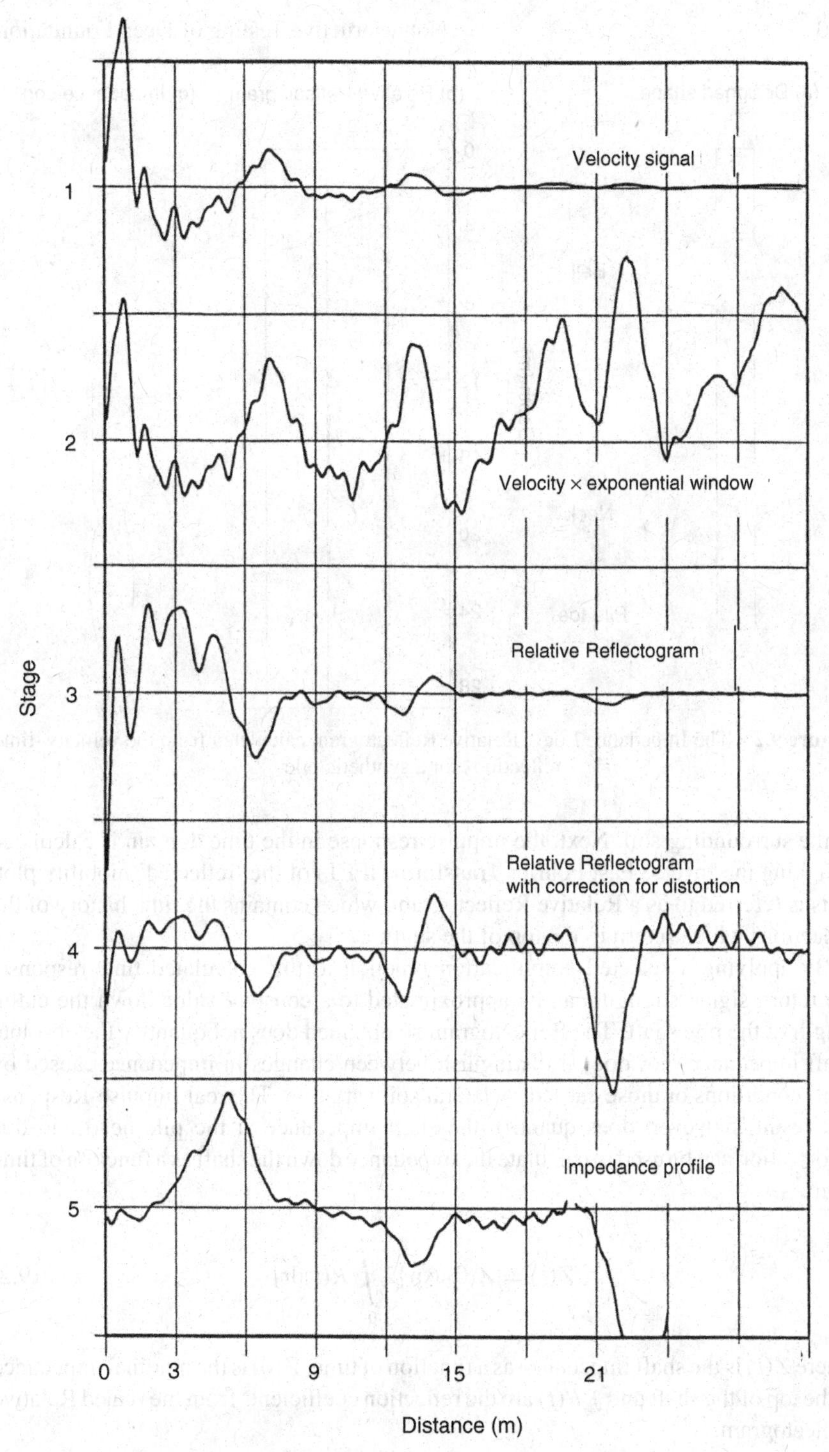

Figure 9.1 Stages involved in generating an impedance profile

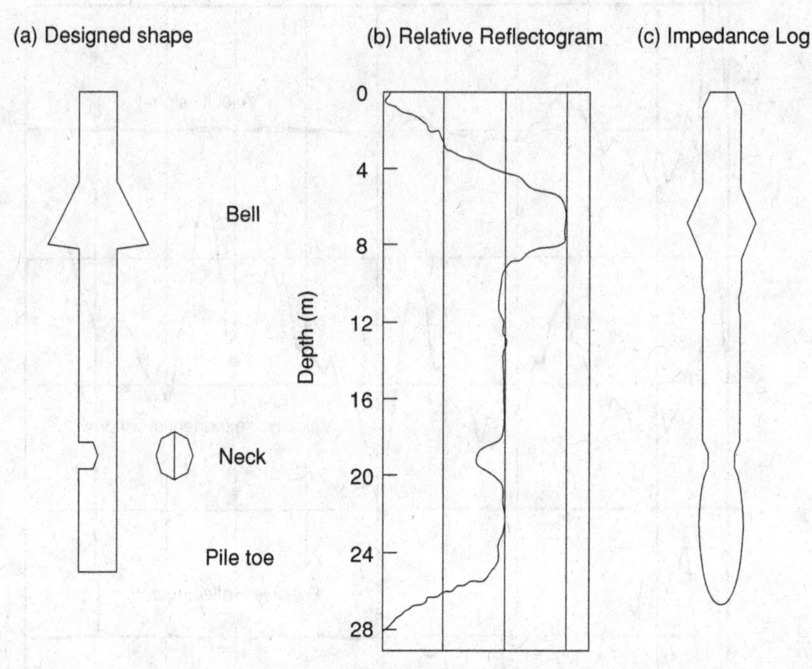

Figure 9.2 The Impedance Log – Relative Reflectogram calculated from the velocity–time reflections for a synthetic pile

in the surrounding soil. Next, the impulse response in the time domain is calculated by taking the Inverse Fast Fourier Transform (IFFT) of the 'reflected' mobility plot. This is referred to as a Relative Reflectogram, which contains the time history of the reflections which return to the top of the shaft.

By applying a selected amplification function to this calculated time-response, the return signal strength can be approximated to a constant value down the entire length of the pile shaft. The Reflectogram so obtained does not quantify the absolute shaft impedance, nor does it distinguish between changes in impedance caused by shaft conditions or those caused by lateral soil variation. The real Impulse-Response test result, however, does quantify the shaft impedance at the pile head and this information can be used to calculate the impedance down the shaft as a function of time from:

$$Z(t) = Z(0)\exp\left[2\int_0^t R(t)\mathrm{d}t\right] \qquad (9.2)$$

where $Z(t)$ is the shaft impedance as a function of time, $Z(0)$ is the nominal impedance at the top of the shaft and $\int R(t)$ are the reflection coefficients from the scaled Relative Reflectogram.

Finally, the impedance as a function of depth is obtained from $Z(t)$ by converting time to depth using the bar-wave velocity of the pile material (Figure 9.1). As in simulation of the Impulse-Response method, if the density and bar-wave velocity of the pile are known, changes in impedance correspond to changes in the cross-sectional area of the pile. The resulting impedance–depth plot can be drawn as a plot of either diameter or cross-sectional area with depth. This is the final Impedance Log, as shown in Figure 9.2. As in other pile–response simulations, the analyst must have all available soil data and pile construction records to accurately calculate the Relative Reflectogram.

The first published Impedance-Log test results appeared in the report on the FHwA study on pile integrity testing (Baker *et al.*, 1993), with presentation of Impedance Logs for the piles tested in San Francisco sands and Texas clays (Figure 9.3).

Very good agreement could be seen between the Impedance-Log profiles and the known shaft shapes in this Class-A exercise. Further Impedance Logs for piles under existing buildings damaged by the 1994 Northridge, California earthquake are given in Davis (1997).

Impedance Logs can be useful in studying the effect of modification to drilled shafts by grouting after construction. As an example, when shaft base defects have been discovered by CSL testing or by coring, it has been possible to improve the

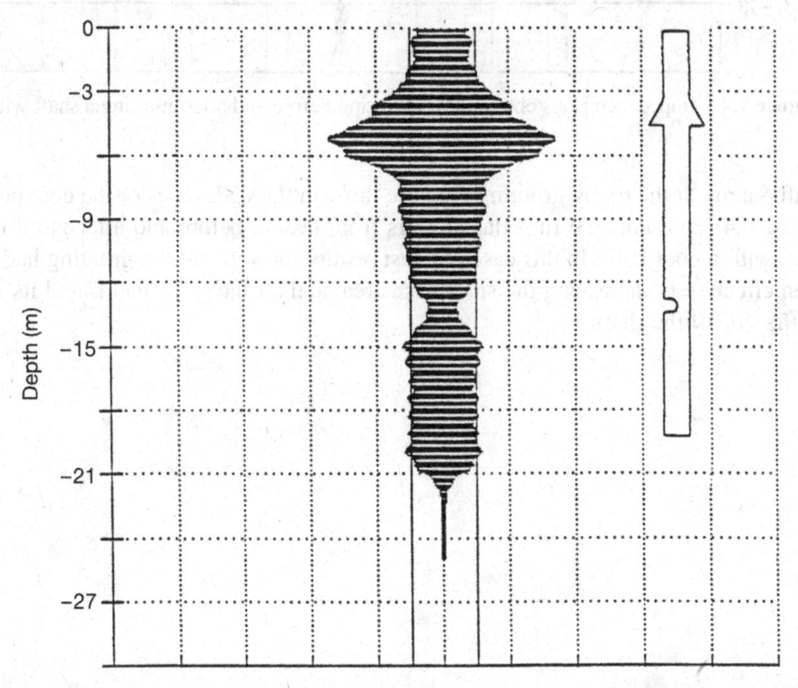

Figure 9.3 Sample of output from an Impedance-Log program – Impedance Log area profile from FHwA trials (San Jose Pile No. 9) (after Baker *et al.*, 1993)

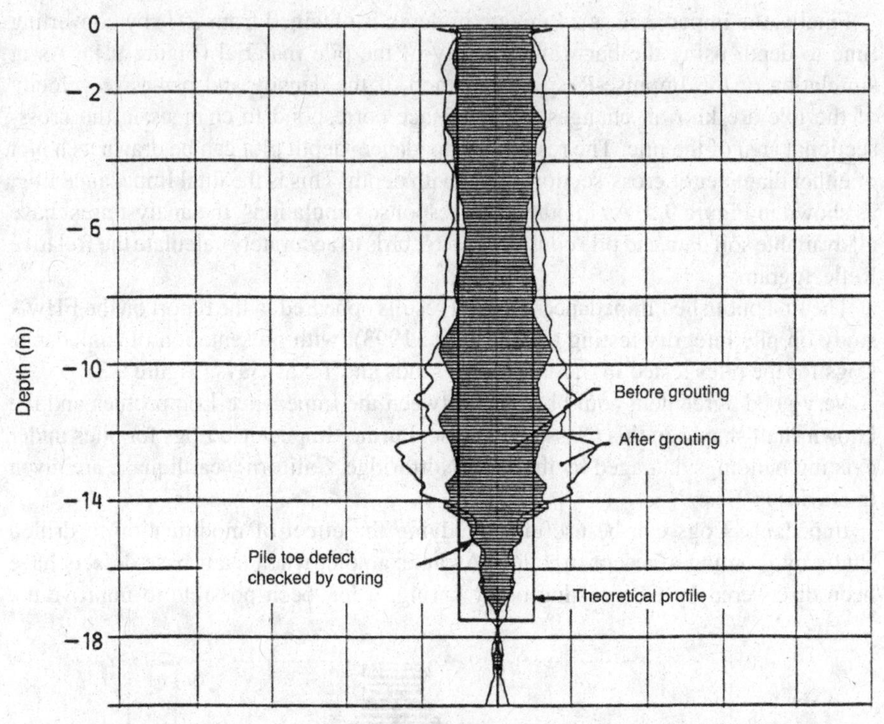

Figure 9.4 Impedance Logs obtained from testing before and after grouting a shaft with a poor base

shaft quality at the tip by grouting the base through the CSL tubes or the core holes. Figure 9.4 represents the Impedance Logs from testing before and after grouting a shaft with a poor base. In this case, the test results showed that the grouting had not been effective in increasing the shaft base area, and all the grout had found its way up the side of the shaft.

10

Low-strain Down-hole Tests

10.1 INTRODUCTION

To overcome some of the limitations of the surface testing techniques described in Chapters 7, 8 and 9, several researchers have focused on the possibilities of down-hole techniques, where length/diameter ratio would not be a problem and where depth could be physically measured by the length of cable lowered down the hole. The results of that research are embodied in three down-hole test techniques:

- Cross-Hole Sonic Logging
- Gamma–Gamma Logging
- Parallel Seismic testing.

10.2 CROSS-HOLE SONIC LOGGING

The Cross-Hole Sonic Logging (CSL) method was developed by CEBTP specifically to overcome the length/diameter limitations of the foregoing methods. The method is a down-hole variation of the Ultrasonic Pulse Velocity (UPV) test. A number of access tubes are attached to the reinforcing cage prior to concrete placement. These tubes are typically 40 mm or 50 mm internal diameter. Either steel or PVC pipe may be used. The tubes are filled with water to provide an acoustic coupling to the test transducers. A transmitter and a receiver transducer are lowered down an adjacent pair of tubes in the concrete to be tested and the transit time of an ultrasonic pulse through the material between the tubes is measured by a data-acquisition system. A continuous series of measurements is made as the probes are raised up the tubes, providing a vertical profile of signal transit time. The vertical movement of the probes

Nondestructive Testing of Deep Foundations B.H. Hertlein and A.G. Davis

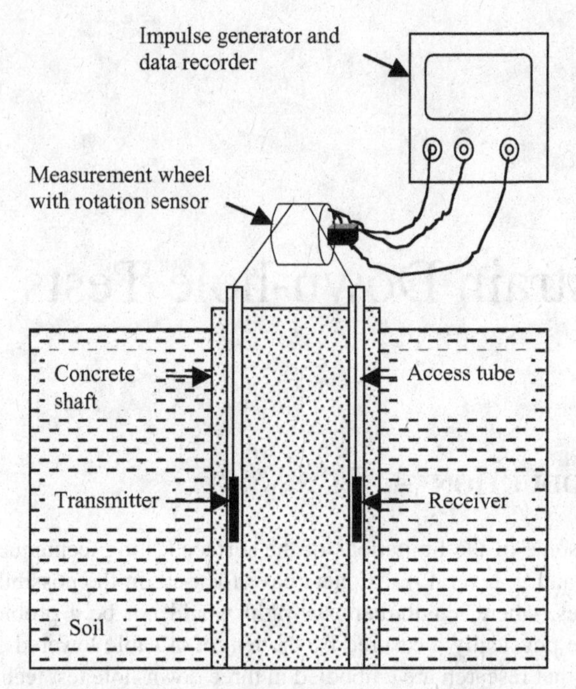

Figure 10.1 Schematic of the Cross-Hole Sonic Log set-up

is measured by a rotation sensor that monitors cable withdrawal, allowing the CSL profile to be directly scaled in depth (Figure 10.1).

10.2.1 CAPABILITIES

UPV is a function of concrete modulus, density and Poisson's ratio, and so the uniformity of the material can be assessed from the uniformity of the CSL profile. Anomalies such as soil inclusions, low-modulus concrete and voids will be readily located by the increase in pulse transit time that they cause. The minimum vertical extent of anomalies that can be detected is determined by the rate of withdrawal of the probes, plus the pulse repetition rate. The minimum detectable horizontal extent of defects depends on the pulse wavelength, or frequency, and the horizontal spacing of the access tubes. Typically, modern equipment operates between 25 and 50 kHz, which allows detection of defects as small as about 2.5 to 4 in in horizontal extent. With current equipment, the maximum recommended horizontal spacing between access tubes is about 12 ft.

The main advantages of CSL are that interpretation is relatively simple, and the depth of the test is limited only by the length of the cables attached to the transmitter and receiver probes. The method is not affected by variations in soil stiffness, or damping characteristics.

Where an anomaly is encountered, multiple tests can be run with the transducers at different depths relative to each other. The resulting combination of pulse paths can then be processed by computer to provide two- or three-dimensional images of the anomaly.

The method has also been successfully applied to submerged structures, or foundations that extend up through water, such as marine jetty piles and dams. The present authors have performed several projects using PVC tubes as transducer guide tubes placed in the water alongside bridge piers in fast-flowing water, dams, nuclear waste storage tanks and fire-damaged oil terminal 'dolphins' in a turbulent ocean environment.

10.2.2 LIMITATIONS AND COST

The limitations of CSL are that fine, horizontal cracks are unlikely to be detected, and 1.5 in minimum id access tubes are usually required for the performance of the test. These can be pre-placed as part of the reinforcing cage, or core-drilled after the concrete is placed. Steel or plastic pipes can be used, but plastic tubes tend to lose their bond to the concrete after a couple of weeks, thus rendering the CSL test ineffective. Plastic tubes therefore limit the time-window available for performing the test, and steel tubes are preferred in most cases.

CSL testing only provides a profile of the concrete between each pair of tubes. If the tubes are attached to the inner face of the reinforcing cage, very little information can be obtained concerning the cover concrete outside of the reinforcing cage.

Derivations of the CEBTP CSL method are really just variations of the way in which the signal data are stored and presented. The CEBTP system modulates the received ultrasonic pulse in such a way that each 'positive-going' peak is printed as a dash whose width matches that of the original peak, and each 'negative peak' is similarly printed as a gap, thus creating a dashed line that contains all of the timing and much of the amplitude information of the original record. Each consecutive pulse is printed contiguously to form a vertical profile, often called a 'waterfall plot', of the data for that particular tube pair (Figure 10.2).

Other versions of the CSL method attempt to assess the uniformity of the concrete by measuring the arrival time of the first peak in the ultrasonic pulse wave train (First Arrival Time, or FAT) and the overall amplitude of the early part of the pulse. The drawback of this method is that the system is basically measuring a relatively small noise, and construction sites are usually noisy! If a computer is trying to detect a peak in a digitized wave-train, it has to be able to differentiate between background noise and the arriving signal. Thus a threshold level has to be set that eliminates the background noise (Figure 10.3). If the first or second peak in the arriving wave-train falls below the threshold level, the computer will latch onto the first peak that exceeds the threshold, be it the second, third or even later arrival, and identify it as the first arrival.

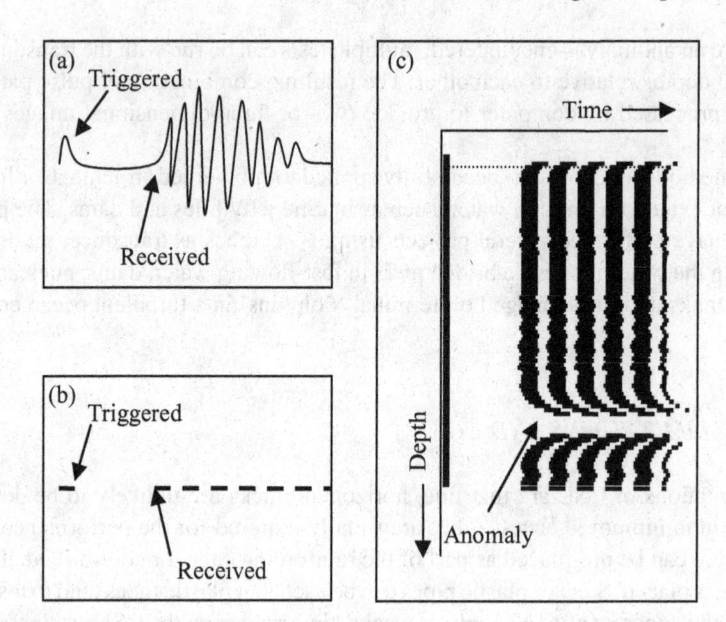

Figure 10.2 Schematic of CSL data reduction and profile compilation: (a) raw pulse data; (b) pulse data modulated to form a single line; (c) modulated pulses stacked to form CSL profile

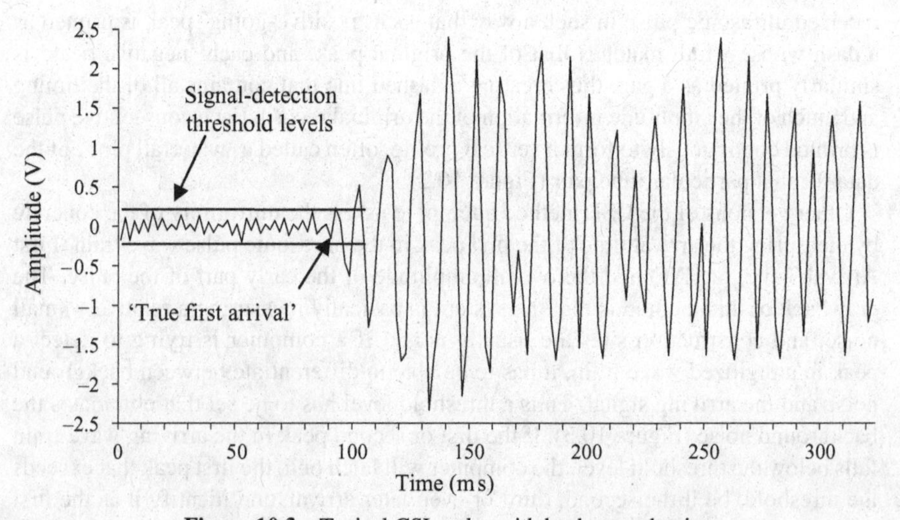

Figure 10.3 Typical CSL pulse with background noise

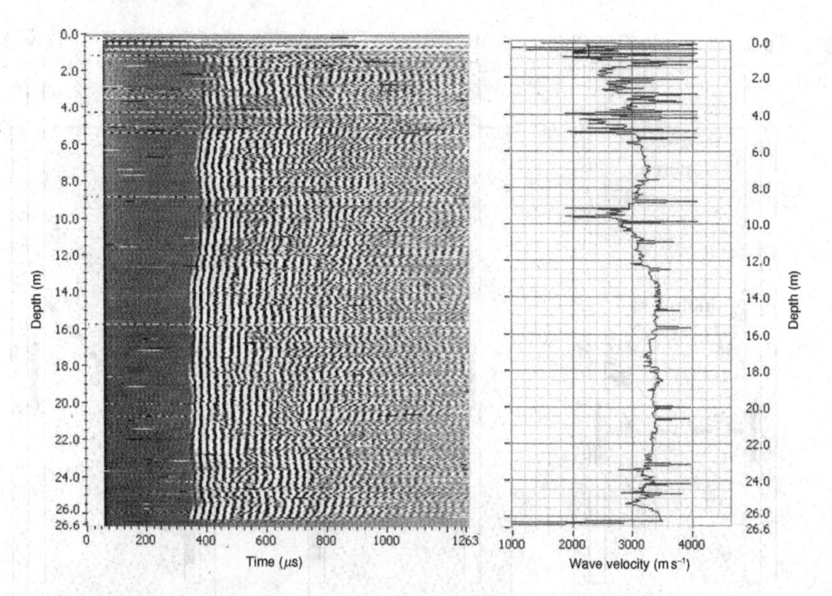

Figure 10.4 Comparison of FAT and 'waterfall' presentation of CSL data

In general the CSL variations which present profiles based on complete pulse time histories ('Waterfall' plots – Figure 10.2) give more consistent output than those which rely entirely on digitization of the data and detection of the FAT, because the coherence of the later and larger peaks in the wave train is often still apparent, even when the first arrival can not be detected (Figure 10.4).

The productivity of CSL testing will depend on the depth of the shafts, and the number of tubes installed. For example, on 48-in shafts with four tubes to a depth of 100 ft, and good access between shafts, a typical testing rate would be about ten to twelve shafts per day.

The CSL test has been standardized by the American Society for Testing and Materials (ASTM) in ASTM D6760, 'Standard Test Method for Integrity Testing of Deep Foundations by Crosshole Testing'. ASTM D6760 has been recognized as an international standard by several countries with no approved standards of their own. The CSL method has also been recognized in some national standards – please refer to Appendix III for additional information.

10.3 CROSS-HOLE TOMOGRAPHY

A significant advantage of CSL over other deep foundation tests is the ability to perform tomographic analysis to create images of an anomaly that show the shape and lateral position of the affected zone. If an anomaly is noted in a normal CSL

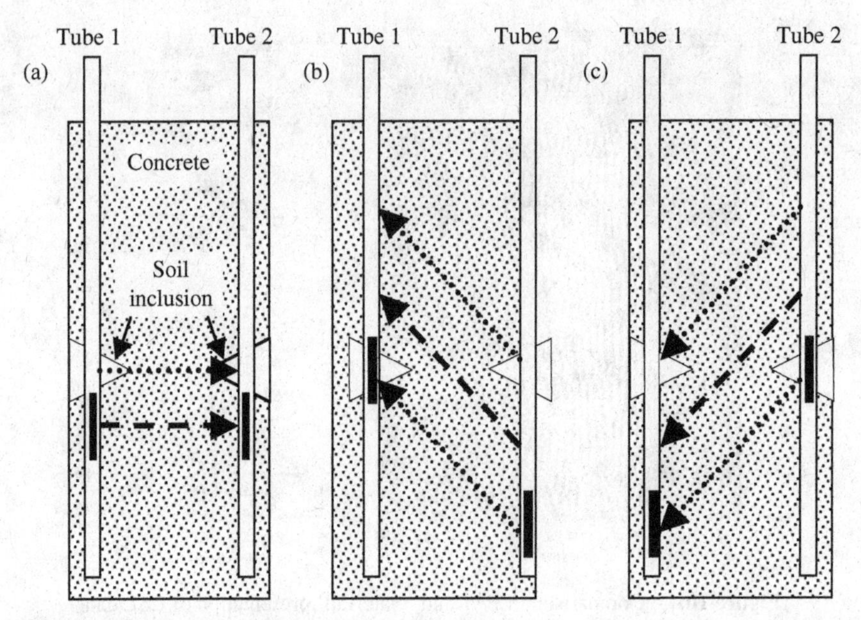

Figure 10.5 Offset CSL tests for tomographic analysis: (a) profile 1–2: transducers at same level – pulse path horizontal; (b) profile 1r–2: transducer in Tube 1 raised – pulse path at 45 °; (c) profile 1–2r: transducer in Tube 2 raised – angle of pulse path reversed

test profile, where the pulse path between the transducers is horizontal, additional profiles are developed with the transducers offset vertically to provide angled pulse paths (Figure 10.5).

A simple way to do this is to raise one transducer by the same amount as the distance between the access tubes, thus creating a 45 ° path between the transducers. Develop the profile, then replace the transducers in the access tubes and reverse the angle of the pulse path by raising the other transducer. The anomalous data from the horizontal test and the two angled tests are then plotted to scale on a simple section drawing of the shaft (Figure 10.6). The anomalous zone(s) will be located where the lines of anomalous data intersect.

When the transducers are withdrawn to develop the profile in the case illustrated in Figure 10.5, the highest transducer in each case will encounter the anomaly first, affecting the pulse arrival time and amplitude. Then there may be a zone where the transducers are above and below the anomalous zone, and the pulse path is unaffected, and finally the lower transducer will encounter the anomalous zone. The angled test data will then show two distinct anomalies where the normal horizontal test data showed only a single zone. This is typical of a peripheral or 'donut-shaped' anomaly Figure 10.6(a). Where all test profiles show a single anomaly, either it extends across the full shaft cross-section, or the vertical extent is so great that a large enough pulse path angle cannot be achieved (Figure 10.6(b)).

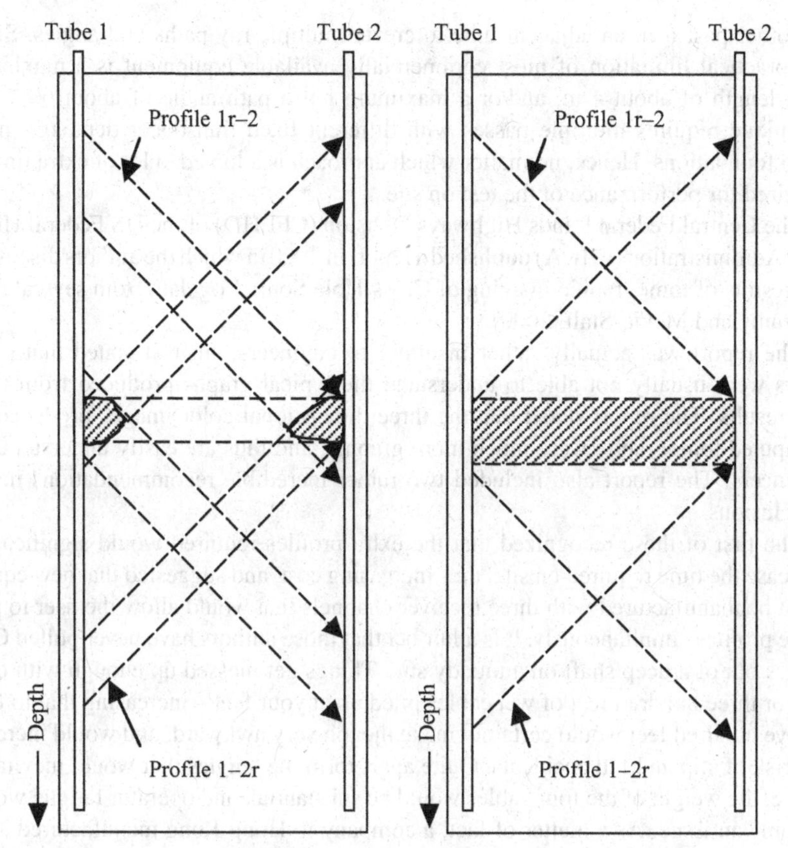

Figure 10.6 Simple tomographic analysis: (a) probable shape of peripheral anomaly; (b) probable lens of anomalous material across entire cross-section of shaft

At the time of writing, several companies are offering software to perform computed tomography, or are providing the data-reduction and plotting service. Some of them are implying that they invented the whole process, either because they have not 'done their homework', or they to not understand the not-so-subtle distinction between 'tomography' and 'computed tomography'. Simple tomography, as described above, has been performed since the late 1960s (Paquet, 1969). The capabilities and limitations of the technique have been well-known for some time, but some modern practitioners are apparently ignoring these well-established facts.

One of the drawbacks of computed tomography is the sensitivity of the data to the angles used for the offset pulse paths. The combination of one horizontal and two opposite-angled offset paths is fine for the simple tomography discussed above, but is not considered adequate for some computed tomography programs. Some recommend at least two different angles in each direction, thus requiring a total of five profiles instead of three. Others prefer to leave one transducer at a fixed depth, and pull the other past it in an adjacent tube to create multiple ray paths and angles. Since the practical limitation of most commercially available equipment is a maximum path length of about 4 m, and/or a maximum pulse path angle of about 60 °, this technique requires multiple passes with different fixed transducer depths in most deep foundations. Hence, no matter which approach is adopted, a lot of extra time is required for performance of the test on site.

The Central Federal Lands Highways Division (CFLHD) of the US Federal Highway Administration (FHwA) published a report in 2000 in which the authors discussed the results of tomographic imaging of Cross-Hole Sonic Log data from several sites (Haramy and Mekic-Stall, 2000)

The report was actually rather insulting to engineers, since it stated that engineers were usually not able to understand the typical graphs produced from CSL test results. The report stated that the three-dimensional color images produced by computed tomography were much more graphic, and thus are easily understood by engineers. The report also included two rather incredible recommendations in the conclusions.

The first of these recognized that the extra profiles required would significantly increase the time required on site, thus increasing cost, and suggested that new equipment be manufactured with three receiver channels that would allow the user to pull three profiles simultaneously. It is a fair bet that those authors have never pulled CSL cables out of a deep shaft on a muddy site. Things get messed up enough with only two or three hundred feet of wet cables piled up at your feet – increasing that to four or five hundred feet would certainly make the job very awkward, and would increase the risk of slip and fall accidents. Quite apart form the tangles that would inevitably ensue, the weight of the four cables would be substantial, and operator fatigue would become an issue. As a matter of fact, a company in Hong Kong manufactured a set of CSL equipment that had three receiver channels and four transducer cables. It did not sell very well, and we do not know of anyone who uses it in multi-channel mode today.

The second recommendation in the FHwA-CFLHD report was that computed tomography should be performed on all CSL tests. This would result in an incredible amount of waste. If the normal CSL profiles show no evidence of anomalies, then neither will the computed tomography – so why increase the amount of work by three or five times if it serves no purpose other than spending tax dollars? Unfortunately, this simple logic somehow eluded 'the powers that be' at the FHwA, because, at the time of writing this chapter, the standard specifications issued by several states, based on the FHwA recommendations for CSL testing, required computed tomography of all CSL profiles. That recommendation appears to have been withdrawn at the time of submission of this completed manuscript, but for the time being, at least, the damage has been done.

Despite the FHwA report – or perhaps because of it – there is some controversy over the value of computed tomography. One of the major issues is the claim of 'increased accuracy' made by several of the software companies. Since computed tomography is performed on CSL data, it is subject to the same factors affecting accuracy:

- The true positions of the access tubes relative to each other and to the body of the shaft are only known at the top of the shaft.
- Computed tomography programs assume that the ultrasonic pulse propagates in a straight line between transmitter and receiver. The actual pulse path is determined by the relative moduli of the concrete particles that it passes through. Snell's law applies – an elastic wave passing through the interface between materials with differing elastic moduli will be refracted – the angle of refraction will depend on the difference between the moduli of the two materials – thus the actual path of an ultrasonic wave usually has an approximately elliptical shape of unknown proportions.
- If the anomaly is caused by the access tube debonding from the concrete, tomography will provide no useful information about that anomaly.
- If the anomaly is at the base of the shaft, angled ray-paths through the anomaly will be minimal, and computed tomography will provide little or no useful information that was not already apparent in the basic CSL profiles.

Thus, the images produced by computed tomography, whether two-dimensional or three-dimensional, are estimates of the shape and size of the anomaly. The casual observer would see a 'picture' and believe that it truly represented the feature, but the smart engineer would be wise to remember that it is only an approximation.

Then there is the cost factor. True three-dimensional (3-D) tomography, in particular, requires a considerable amount of time to perform, both in the field and in the post-processing and is thus usually more costly than CSL with simple tomography. Some software and equipment manufacturers have attempted to reduce the cost of performing computed tomography by eliminating the angled pulse paths. In a paper published at the 2003 NDT-CE symposium in Berlin, Germany, Vólkovoy and Stain describe an analysis algorithm that takes the conventional horizontally oriented CSL data and generates a series of horizontal slices through the shaft (Volkovoy and Stain, 2003). Since only horizontally oriented data are used in the analysis, it is not possible to determine the actual horizontal extent or true location of an anomaly – only that the pulse path between the tubes in question has been affected. A smoothing process converts the images from adjacent layers into a 3-D rendering, but it is an approximation, not an actual image of the shape and location of the anomaly. At least one other commercially available tomographic analysis program uses a similar procedure, despite the seller's claims of enhanced accuracy – let the buyer beware!

Whichever method is used to generate the 3-D images, the end result is usually presented in a written report with two-dimensional (2-D) graphics. This raises questions as to the cost-effectiveness of 3-D modeling. Whether the estimated shape of an anomaly shown in a 2-D rendering of a 3-D model provides extra information that is worth the extra cost is up to the individual user to decide. The present authors can

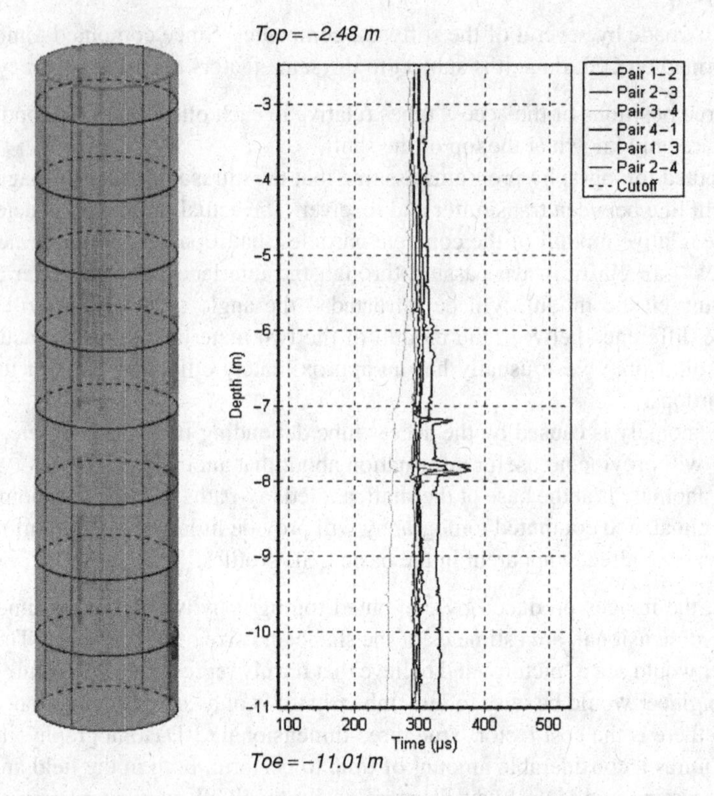

Figure 10.7 Typical three-dimensional tomography image – side view. Reproduced by permission of Testconsult Ltd, Risley, UK (note: original in colour)

only recommend that the user considers the limitations of the various techniques, and sets his or her expectations at a reasonable level!

Figures 10.7 and 10.8 present typical three-dimensional tomography images, showing a side view and a horizontal slice, respectively.

10.3.1 SAMPLE SPECIFICATION

For the reader who is considering the specification of nondestructive integrity tests for deep foundations, a sample specification for the Cross-Hole Sonic Log method is given in Appendix IV of this book.

10.4 SINGLE-HOLE SONIC LOGGING

Much recent publicity has been given to a derivation of the CSL method which some practitioners claim to have developed specifically for augered-cast-in-place

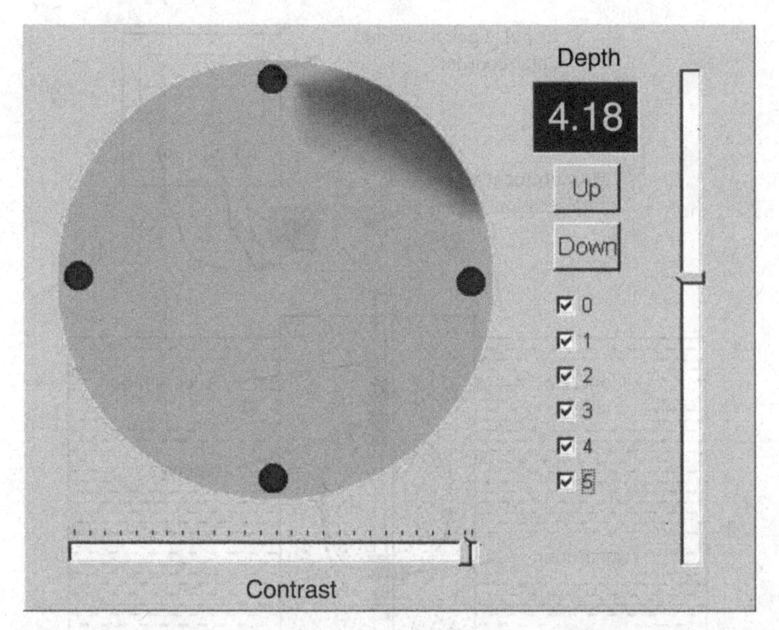

Figure 10.8 Typical three-dimensional tomography image – horizontal slice. Reproduced by permission of Testconsult Ltd, Risley, UK (note: original in colour)

(ACIP) piles, where a single tube is installed in the pile by placement in the auger stem prior to drilling, or is pushed down into the grout after completion of the pile. Both the receiver and transmitter probes are placed in the single tube, a certain distance apart vertically. This method is known as Single-Hole Sonic Logging (SSL) (Figure 10.9). In fact, Weltman described single-tube testing in his report to CIRIA in 1977 (Weltman, 1977).

10.4.1 CAPABILITIES

In grout shafts, SSL can be used to assess the uniformity of grout quality before it is fully hardened. In the event of an anomaly being detected, this affords the contractor the opportunity of re-drilling the shaft while the grout is still soft and relatively easy to remove (Brettman, *et al.*, 1996).

In the case of CSL tests where it is suspected that one or more of the access tubes may have suffered a loss of bond to the concrete, SSL tests can sometimes be used to confirm whether debonding has occurred or not.

10.4.2 LIMITATIONS

At least one access tube must be placed in the shaft for the SSL test. The use of steel access tubes, while preferable for CSL, is not recommended for SSL. Most

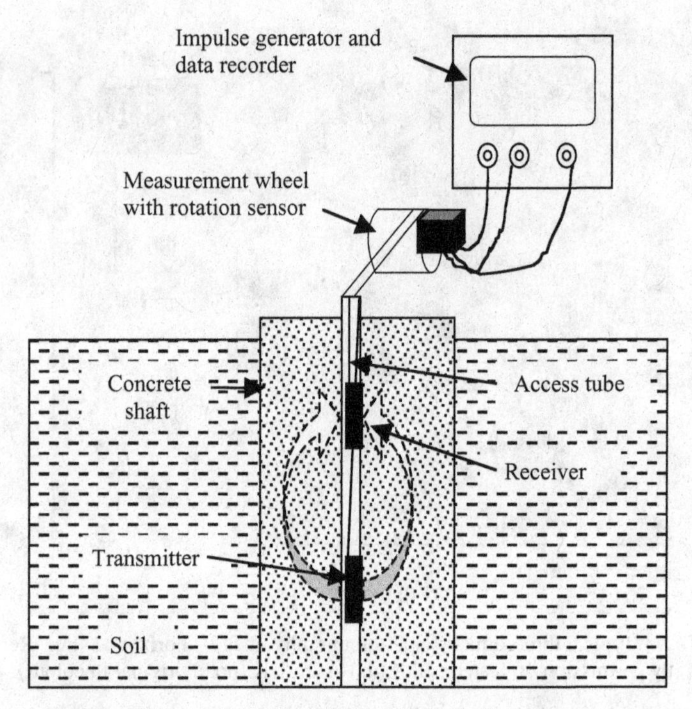

Figure 10.9 Schematic of the Single-Hole Sonic Log set-up

physics majors and all geophysicists are familiar with Snell's law of refraction, but it is not so well-known among civil engineers. When a wave-train passes through the interface between two different materials at an angle, the wavepath will change, depending on the difference in wave velocities of the materials. Snell's law is a way of calculating that amount by which the wave path changes, but for the purposes of this book it is enough to understand that a wave passing from a high-velocity material into a lower-velocity material will bend towards the normal (perpendicular to the plane of the interface), whereas a wave passing from a low-velocity material into a higher-velocity material will bend away from the normal (i.e. closer to the plane of the interface). Research performed by PileTest in Israel has shown that the angle of refraction caused by the difference in modulus, hence pulse velocity, between the steel and the concrete results in most of the pulse energy radiating away from the tube. The lower modulus of the PVC results in most of the energy being refracted almost parallel to the tube (Figure 10.10). SSL is therefore much more effective in PVC access tubes (Amir, 2001).

If the PVC pipe is to be inserted after grout placement, it usually requires additional support or stiffening. For shorter lengths, it may be practical to place a length of reinforcing steel inside the PVC during insertion, and then remove the steel after the PVC pipe is in position. For longer shafts, this may be impractical, and a length of

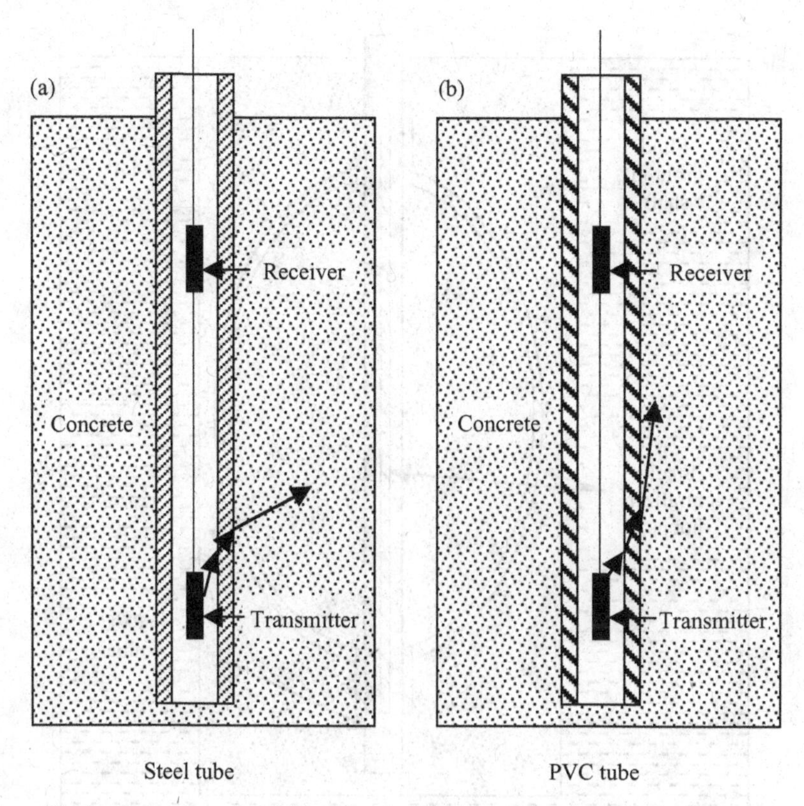

Figure 10.10 Refraction of an SSL pulse at a tube/concrete interface – Snell's law: (a) steel tube; (b) PVC tube with permission from ASCE

reinforcing steel may be tied to the outside of the PVC. Sometimes a centralizer or 'basket' of lighter reinforcing steel may be tied to the bottom of the PVC Pipe. If the centralizer is appropriately sized, it becomes an inspection tool in itself, since any squeezing or necking of the shaft will be felt as the centralizer tries to pass through that zone (Figure 10.11).

SSL productivity will depend on site access conditions and pile length, but as an example, with a pile length of 80 ft and good access between piles, it is possible to test more than fifty piles per day.

10.5 GAMMA–GAMMA LOGGING

Gamma Logging of concrete foundation shafts was developed from borehole geophysics. The original method, passive Gamma Logging, measured the natural gamma radiation of the soil strata. For foundation testing, a gamma emission source is included

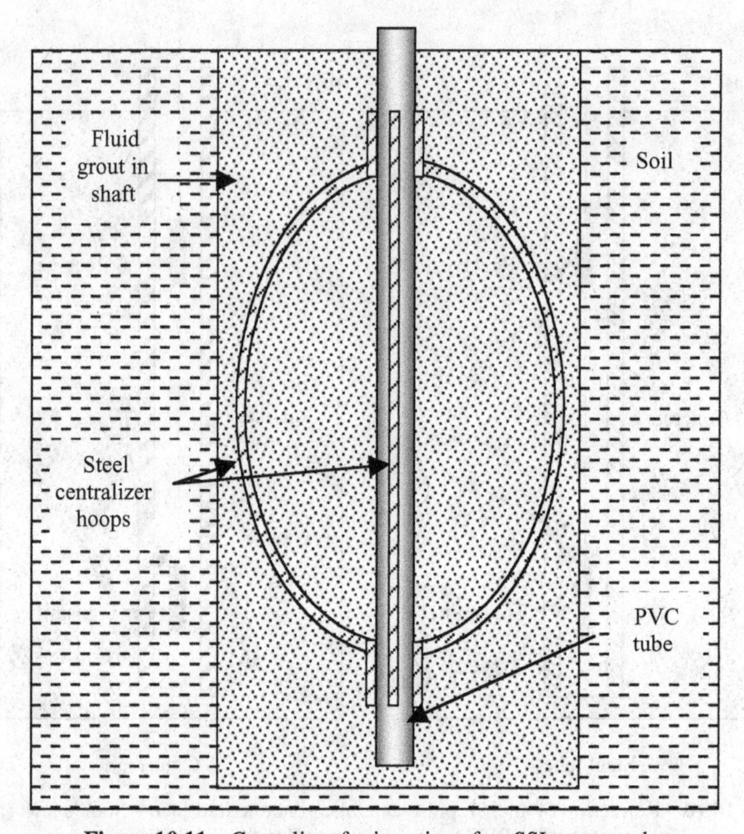

Figure 10.11 Centralizer for insertion of an SSL access tube

in the test probe, and the detector measures the amount of gamma radiation, both natural and introduced, that is returned to the probe. It is therefore more correctly referred to as Gamma–Gamma Logging (GGL).

In this method, a low-energy radioactive source, typically in the range of 10 to 100 mCi, is mounted in the test probe, and emits gamma radiation into the surrounding concrete. Through a phenomenon known as 'Compton scattering', the radiated particles are variously absorbed, refracted or reflected by the surrounding material. Some of the refracted and/or reflected particles eventually make their way to a radiation detector, or scintillometer, in the test probe, where each particle is counted as it passes through the detector. These particles are commonly referred to as 'backscatter'. Concrete absorbs a quantity of the radiated particles in proportion to its density, and therefore affects the number of particles that can be detected by the scintillometer. Since the amount of backscatter detected in a concrete shaft is determined by the density of the surrounding concrete, GGL is effective in assessing concrete uniformity (Preiss and Caiserman, 1975).

For GGL, access tubes are required and they are installed in a similar manner to the access tubes for CSL. It has been found, however, that PVC tubes are better than steel tubes for GGL testing. Steel absorbs so much of the radiation that the sensitivity of the test is significantly reduced. GGL is not sensitive to the bond between the tube and the concrete, and so the loss of bond that is generally experienced with PVC tubes does not reduce the effectiveness of the GGL test.

The transducer probe is lowered to the base of each access tube, and a measurement is made. With older models of Gamma–Gamma equipment, the probe is then raised in uniform increments and a measurement is made at each point. Typically, the probe must remain static for some 15 or 30 s to complete a measurement. The vertical extent of defect or anomaly that can be detected by this form of GGL will depend on the distance between test positions. The current generation of GGL systems is capable of a higher measurement rate, and thus the transducer can be withdrawn continuously via an electrically powered measurement winch. Typically the rate of withdrawal is about 10 ft/min, and measurements are made every 0.1 ft (Figure 10.12).

The data are usually plotted in graph form as a vertical profile of comparative density against depth, where any significant reduction in density will be readily apparent as an inflexion in the graph. Often, the data from multiple tubes may be plotted on the same graph, since significant reductions in density are usually readily apparent. The analysis

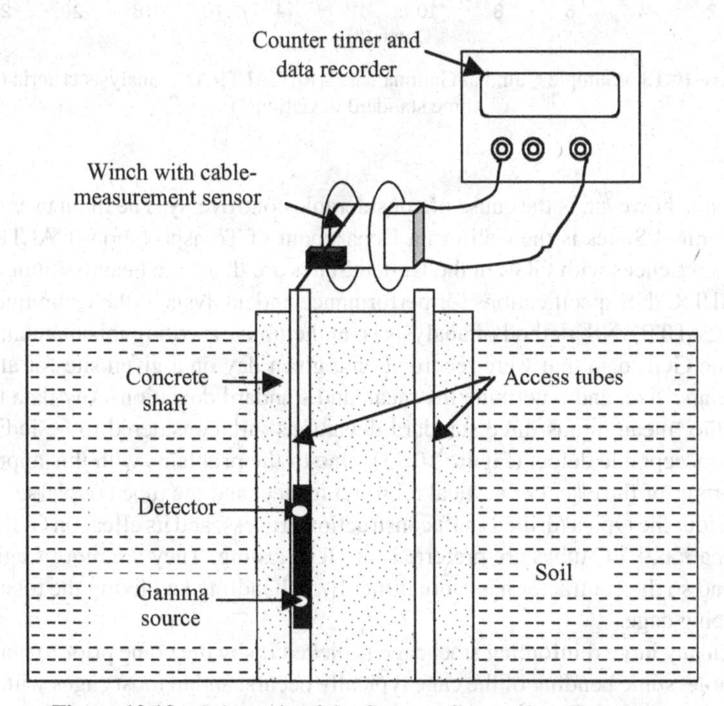

Figure 10.12 Schematic of the Gamma–Gamma Log Test set-up

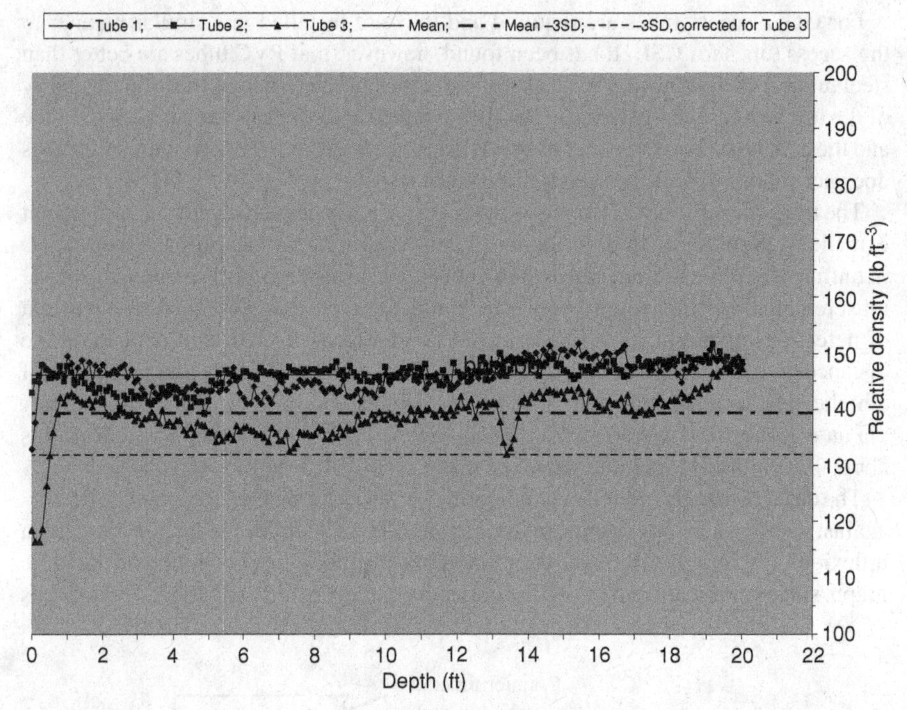

Figure 10.13 Sample Gamma–Gamma data with CALTRANS analysis criteria (3SD, 'three standard deviations')

of the data, however, is the cause of considerable controversy. The main user of GGL in the United States is the California Department of Transportation (CALTRANS). Most experiences with GGL in the United States are therefore heavily influenced by the CALTRANS specifications for performance and analysis of the technique.

The CALTRANS method of analysis is, at the time of writing this manual, to take all of the GGL data that were recorded on a given day on a given site for all shafts of the same size, and determine the mean and standard deviation. Any data that fall below the 'mean minus three standard deviations' are considered to be indications of an unacceptable defect (Figure 10.13). One of the problems with this approach is accounting for the effect of vertical reinforcing steel and the tube couplers.

Consider the typical drilled shaft construction process, and its effect on GGL access tube location. PVC tubes are preferred for GGL testing. They are more fragile than steel, and so the contractor has more difficulty in handling and fixing the tubes to the reinforcing cage.

When any large reinforcing steel cage is picked up by the crane prior to placement in the hole, some bending of the cage typically occurs, and in most cages with helical reinforcing steel, some wracking or 'unwinding' of the cage occurs, resulting in twisting of the GGL access tubes. If the cage is resting on the bottom of the hole when placed, it will tend to rewind to a condition close to what existed prior to being 'picked'. The access tubes, however, may slip and not return to their original positions. The tubes may be further moved by the flow of concrete, particularly in deep, large-diameter shafts. These authors' experience with many years of Cross-Hole Sonic Logging has shown that it is not unusual for access tubes to wander several inches laterally around the cage in either direction. As a result, the proximity of the vertical reinforcing steel is usually not known with any certainty.

Reinforcing steel has a higher density than concrete. If a vertical steel bar is close to a GGL access tube, the apparent recorded density will be higher than for a tube that is not close to a vertical steel bar. Similarly, a tube with a lower thickness of cover concrete than its neighbor is likely to produce lower apparent density measurements, because the sphere of influence of the Gamma-Logging tool will include some of the soil outside the shaft (Figure 10.14).

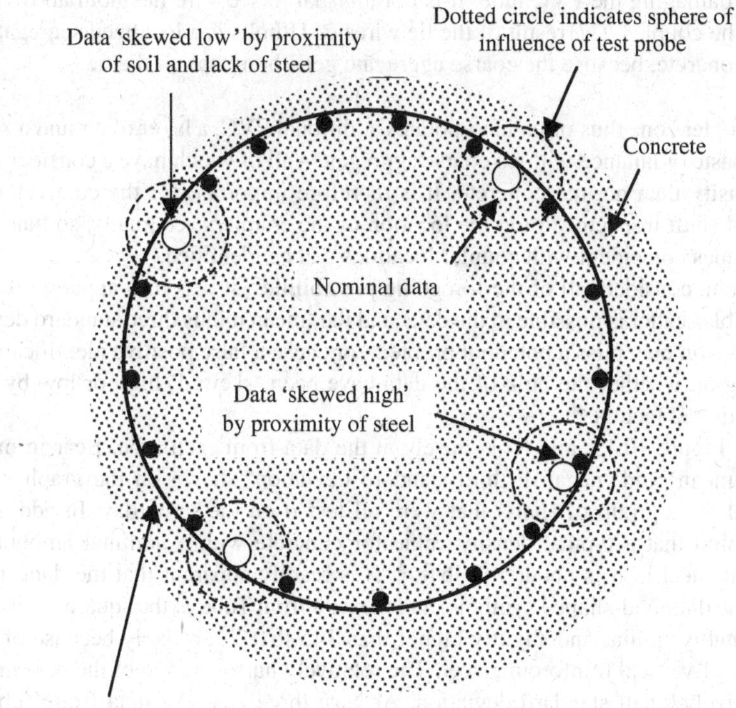

Figure 10.14 How the access tube and reinforcing steel movements can influence apparent density measurements

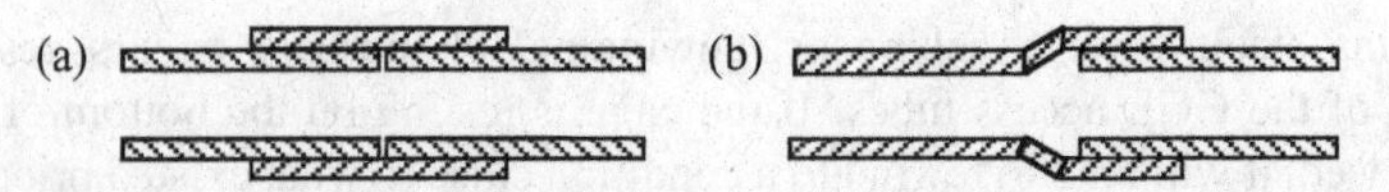

Figure 10.15 Typical couplers used for GGL access tubes: (a) PVC pipe external sleeve joint; (b) PVC pipe 'bell-and spigot' joint

In addition to the foregoing, the typical couplers specified for the access tubes are either external sleeve or bell-and-spigot joints (Figure 10.15), bonded with adhesive. Several factors come into play here:

- There is the additional comparatively low-density mass of the sleeve coupler itself.
- Incomplete application of adhesive can result in a thin layer of air in the joint.
- The coupler sleeve projects out from the tube wall, which makes it an attractive location to place tie wires or U-bolts to fix the tube to the reinforcing cage and so increase resistance to slipping without having to make the fixing so tight that it risks damaging the PVC tube. It is not unusual to see wire ties both above and below the coupler. The result of the tie wires or U-bolts is a localized segregation of the concrete, because the coarse aggregate gets 'hung up' on the tie.

The coupler zone thus often incorporates additional PVC, a layer of air and a zone of grout paste or laitance with no coarse aggregate – all of which have a considerably lower density than normal concrete. It is therefore important for the contractor or the drilled shaft inspector to log the location of the couplers accurately, so that low density zones coincident with coupler locations can be identified.

When one considers all of the foregoing possibilities, it becomes apparent that a considerable scatter is possible in the data, and the 'mean minus three standard deviations' criterion may miss a potential problem, or cause a false positive identification, depending on whether the bulk of the data have been 'skewed' high or low by the aforementioned factors.

Revisit Figure 10.13, and look closely at the data from each tube. Bear in mind that the 'mean' and 'negative third standard deviation' shown on the graph were calculated for a group of shafts that were all tested on the same day. In addition, bear in mind that a typical density value for concrete with a nominal amount of reinforcing steel in it is about 145 lb ft^{-3}. It is readily apparent that the data from Tube 1 (the diamond-shaped symbols) and the data from Tube 2 (the square symbols) are reasonably similar and slightly higher than 145 lb ft^{-3}, probably because of the proximity of vertical reinforcing steel. The relatively narrow range of the data gives us a relatively small standard deviation. At least three zones of data from Tube 3 (the triangular symbols) fall to the left of the 'negative third standard deviation' line, and are thus unacceptable. A close look at the data from Tube 3, however, shows that almost the entire length of the tube produced density data that were appreciably lower those recorded in the other two tubes, probably because of a lack of vertical reinforcing steel in the proximity of the tube.

In cases like this, it would be appropriate to calculate a separate mean for the tube in which the data were 'skewed' low. The wider range of data from this tube, however, would also create a larger standard deviation that would be statistically unrepresentative of the bulk of the data from the other tubes. The conservative approach would be to calculate a mean for the 'skewed' data set, and apply the standard deviation calculated for the group as a whole. Applying these 'corrected' criteria to the data presented in Figure 10.13 then shows that the lower zone of 'lower-density material' barely exceeds the acceptance criteria, and is, in fact, typical of the response caused by the coupler, as discussed earlier. The middle zone now falls within the acceptable range, and only the zone at the top of the shaft is now considered questionable.

CALTRANS have now recognized the validity of some of these concerns, and have recently embarked on a major revision of their analysis procedure, but, at the time of writing this book, there is little information available in the public domain, and only a handful of personal presentations have been made by CALTRANS personnel regarding the new procedures.

10.5.1 CAPABILITIES

Gamma–Gamma Logging allows assessment of the relative density of concrete or grout in a shaft immediately after placement. In the event of an anomaly being detected, this gives the contractor the opportunity to re-drill the shaft while the material is still fluid or soft and easy to remove.

The depth of the probe at each measurement position is known from the control cable length, and the data are plotted as a graph of relative density against depth. Laboratory density measurements and calibration of the probe on a control block made of the same concrete as the shaft to be tested can provide a correlation factor which allows the true density to be calculated from the relative measurements.

With appropriately placed access tubes, the Gamma–Gamma Log can provide information on the quality of the cover concrete, or concrete external to the reinforcing cage.

10.5.2 LIMITATIONS AND COST

Similarly to CSL, Gamma–Gamma Logging requires advance planning because access tubes must be placed or core-drilled in the concrete to be tested. Where pipe is pre-placed, it should be 2.0 or 3.0 in-id-schedule 40 PVC or similar material with a density substantially lower than concrete.

The main limitations of the method are the range of gamma radiation in concrete (typically 3 to 4 in) and the regulations governing use and transportation of radioactive materials. In the United States, the gamma source for GGL equipment can only be transported and used by personnel trained and licensed according to Nuclear Regulatory Commission requirements. The actual licensing requirements vary from state to state, and so transporting the equipment across state lines can require time-consuming paperwork and substantial temporary license fees. Similar restrictions apply in several other countries.

Using the same number of access tubes, GGL can provide better quantification of material density than CSL, but covers much less of the concrete cross-section within any given shaft (Davis and Hertlein, 1994).

The productivity of GGL is dependent on tube length and quantity, and the type of equipment used. It is therefore very difficult to generalize either productivity or cost, but for budgetary purposes, the method can be considered to be approximately 50 % more costly than CSL testing.

There was a surge of interest in the method in the early 1980s, but the liabilities involved in transporting and using a radioactive source in deep foundations are substantial. Experience with leaking radiation sources and getting probes jammed in access tubes demonstrated the significant financial liability incurred when sites had to be cleaned up.

10.6 PARALLEL SEISMIC TESTING

To be most effective, all of the foregoing methods require access to the head of the shaft, and so are best suited to new construction. The Parallel Seismic method was developed by the CEBTP specifically for existing structures, or situations where the head of the shaft to be tested was not accessible.

In order to perform the test, a borehole is drilled in the soil close to, and parallel with, the foundation to be tested. The borehole is lined with 40 mm-id-schedule 40 or similar PVC pipe, which is filled with water to provide acoustic coupling. A hydrophone receiver and a hammer with an integral trigger device are connected to a data-acquisition system.

The hydrophone is lowered down the borehole to the base, and the structure close to the top of the tube is struck with the hammer. An acoustic pulse is generated by the hammer blow, which also starts the data-acquisition cycle. The hydrophone signal is recorded, allowing the time of arrival of the acoustic wave at the hydrophone to be determined. If there is a high level of background noise on the site, or if the arriving signal is attenuated by distance or soil conditions, the hammer impulse may be repeated several times, and the resulting signals stacked (summed, then averaged) to reduce the random noise and enhance the coherent signal. The hydrophone is raised in uniform increments and the test is repeated at each increment (Figure 10.16).

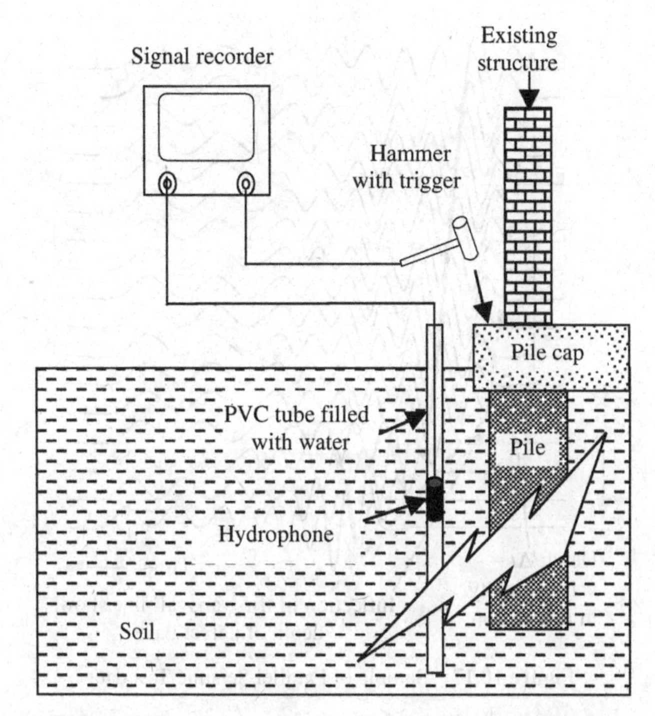

Figure 10.16 Schematic of the Parallel Seismic Test set-up

If the distance between the access tube and the foundation is assumed to be relatively constant, then the primary effect on signal transit time will be the length of the foundation through which the signal has passed. Since the depth of the transducer changes in uniform increments, so the transit time will change proportionately while the probe is alongside the foundation. The velocity of the signal in soil is considerably lower than through steel, concrete or timber. Where the signal encounters a defect or the end of the shaft, the path length around the defect or through additional soil to the receiver will cause a greater increase in transit time at that point.

10.6.1 CAPABILITIES

The output of the test is a stacked graph of time against depth, in which each hydrophone trace is plotted in sequence. Where the signal has passed through a sound, continuous foundation, a line drawn on the graph to link first-wave arrival points will show a uniform slope. This slope is determined by the wave velocity through the foundation material. Since each type of foundation material has a characteristic range of velocity, it is possible to determine from the slope the material from which an unknown foundation is constructed.

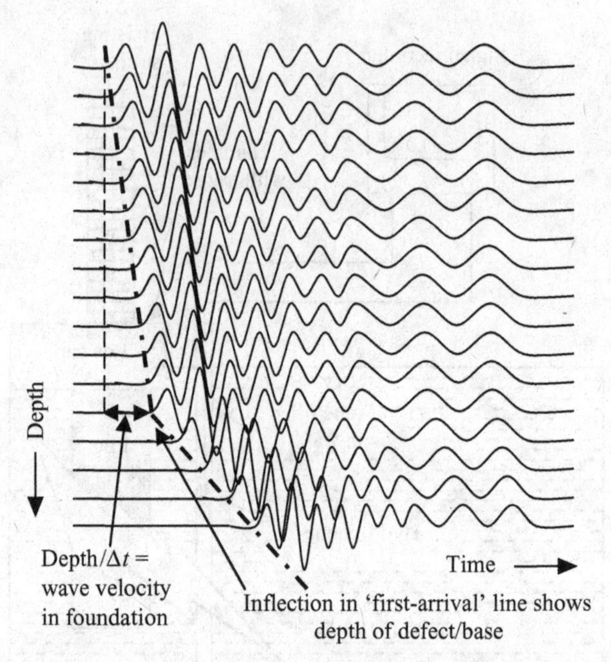

Figure 10.17 Sample of Parallel Seismic Test data

Where transit time is increased by a defect or the additional soil below the base of the shaft, an inflection will be apparent in the line linking 'first arrival points' (Figure 10.17). Since the vertical axis is directly scaled in depth, a minimum of interpretation skill is required for the Parallel Seismic method.

10.6.2 LIMITATIONS AND COST

The limitations of the method are that a borehole must be drilled to provide access for the hydrophone, and typically the access tube must be within 3 ft or so of the foundation to be tested. Alternatives to drilled boreholes have been successfully utilized. In 1984, a large project was conducted in London, UK, where a Terrahammer 'Thrustmole' was used to place access tubes depths of more than 25 ft, through fill and rubble into the underlying clay.

Fill and rubble can be a problem. The soil between the foundation and the access tube must be capable of propagating a low-strain acoustic impulse. The method will not work in dry or loose granular material, or unconsolidated fill.

The impact point on the structure must be capable of sustaining the impact of the hammer without damage. The impact point must have good mechanical coupling to the top of the foundation, and should be as close as possible to the axis of the shaft. For example, delivering the impact to a grade beam several feet away from the foundation shaft is unlikely to produce a clear and repeatable wave in the shaft.

Productivity and cost are very dependent on site conditions and the required depth of the test, since poor soil conditions may require the acquisition and stacking of multiple recordings at each measurement position. In the present authors' experience on typical projects, the costs are similar to CSL testing.

11

Field Mock-ups of Deep Foundations: Class-A Predictions

Pile integrity testing of deep foundation shafts is difficult to reproduce in controlled laboratory conditions. Cross-Hole Sonic and down-hole Gamma–Gamma Logging techniques can be modeled in the laboratory, as described by Stain and Williams (1991). However, shaft head impact tests rely upon the channeling of dispersive stress waves down cylindrical or prismatic structures with length/diameter ratios typically between 10:1 and 40:1. Paquet (1968) points out that the frequencies used for both harmonic (Vibration and Impulse Response) and Sonic-Echo tests are relatively low, and the corresponding wavelengths are great compared with typical pile diameters.

Both the Impulse-Response and Sonic-Echo tests are affected by soil damping. If it is required to study the effect of soil damping on the shaft response, then the wavelengths in the soil must be much greater than the shaft diameter. Take, for example, a 1.5 m diameter shaft in a soil with a shear-wave velocity of 800 m/s. At a frequency of 500 Hz, the wavelength is 1.6 m, close to the pile diameter. At higher frequencies, the calculated damping would reduce to a poor approximation.

For these reasons, tests on scaled-down laboratory foundation models do not produce satisfactory results. This was realized early on in the development of these test methods and several full-scale test sites have been constructed worldwide to test the validity and accuracy of the different methods. Usually, various defects or shape changes have been included in the test shafts. These full-scale experimental sites include the following:

- London, UK, 1969 (Levy, 1970)
- Blyth, Northumberland, UK, 1985 (Lilley et al., 1987; Kilkenny *et al.*, 1988)
- Ghent, Belgium, 1987 (Holeyman *et al.*, 1988)
- FHWA trials, USA, 1989–1991 (Baker *et al.*, 1993)

Nondestructive Testing of Deep Foundations B.H. Hertlein and A.G. Davis

- Cupertino and San Jose, California, USA, 1989
- Bryan, Texas, USA, 1990
- Delft, Netherlands, 1991 (Wheeler, 1992)
- Houston, Texas, USA, 1996 (Samman and O'Neill, 1997a, b)
- Amherst, Massachusetts, USA, 2000 (Iskander *et al.*, 2001).

In several of these cases, initial tests were performed 'blind', that is, without prior knowledge of the lengths of the shafts or the location and type of built-in defects and shape changes. This type of exercise is referred to as a 'Class-A Prediction'. The sponsors of the test sites then assessed the accuracy and reliability of the test methods. At least three of the sites have been preserved for future testing as methods are refined and developed (Blyth, Northumberland, UK, Bryan, Texas and Amherst, Massachusetts, USA).

The first study in 1969 was organized by the Greater London Council to check the validity of the CSL method for detecting necking and included voids in bored piles. This program was initiated because of the increasing size of bored piles being used in the London area, carrying loads up to 1000 tonnes. Any potential defects in these larger piles were considered to have more serious consequences than for the smaller-diameter piles in use up to that time. Four 480-mm-diameter piles were constructed, one with no built-in defects and three with a variety of simulated defects such as sand/gravel layers, clay inclusions and voids. Three steel tubes were cast in the piles for CSL testing. The conclusion of the organizers was that the CSL test was very successful in locating significant defects, including the detection of poorer concrete at the pile toe that was not planned.

Piles with planned defects were constructed between 1970 and 1985 for testing NDT methods, but in all of these cases independent assessors did not referee the test programs. With the growth in commercial use of the methods as quality control in new construction throughout the world and the increasing tendency for engineers to rely on the test results to accept or reject piles, engineering communities in the 1980s considered that independent assessment of the various methods available was becoming a pressing need.

The three studies set up in the 1980s (UK, Belgium and FHwA Erials, USA) were supported by the respective National Societies of Soil Mechanics, and the local piling industry donated the time and materials for each project. In two of the cases (Bryan, Texas, USA and Blyth, Northumberland, UK) these were designated as 'National Research Sites'. In each case, companies and organizations specializing in the different nondestructive methods were invited to test the piles, without knowledge of any built-in defects. The various sites had different pile sizes and soil conditions, as follows.

Blyth, Northumberland, UK (Kilkenny *et al.*, 1988)

- Soil description: glacial till (stiff to very stiff clay) with typical undrained shear strength of 160 kN/m^2.
- Number of piles: 25.

- Pile diameter: 750 mm.
- Pile lengths: 11–22 m.
- 12 of the piles were equipped with CSL tubes.

Ghent, Belgium (Holeyman *et al.*, 1988)

- Soil description: silty sand (medium-to-dense), with typical Standard Penetration Test (SPT) numbers (*N*) of 10–15 for the top 10 m and > 35 below that depth.
- Number of piles: 5 pre-cast square-section piles, 10 screw piles and 5 cast-in-place piles.
- Pile diameters: 350–450 mm.
- Pile lengths:13–14.5 m.
- No CLS tubes.

San Jose, California, USA, FHwA study, 1989–1991 (Baker *et al.*, 1993)

- Soil description: stiff sandy clay, 0–1 m; soft-to-medium silty clays, 1–15 m; stiff silty clay, 15–22 m.
- Number of piles: 6 cast-in-place drilled shafts; variety of slurry drilling techniques.
- Pile diameter: 900 mm.
- Pile lengths: 2 of 9 m and 4 of 17.5–18 m.
- All piles equipped with CSL and GGL tubes.

Cupertino, California, USA, FHwA study, 1989–1991 (Baker *et al.*, 1993)

- Soil description: dense sandy clay, 0–1 m; dense clayey and sandy gravels, 1–10 m; very dense silty and clean sands, 10–13 m.
- Number of piles: 5 cast-in-place drilled shafts, cast in dry open holes.
- Pile diameter: 900 mm.
- Pile lengths: 7–9 m.
- All piles equipped with CSL and GGL tubes.

Bryan, Texas, USA (1), FHwA study, 1989–1991 (Baker *et al.*, 1993)

- Sand site.
- Soil description: loose becoming medium-dense fine sand to 14 m below ground level, with Standard Penetration Test (SPT) numbers (*N*) of 8–30; hard clay below 14 m.
- Number of piles: 5 cast-in-place drilled shafts; water/bentonite with tremie concrete placement.
- Pile diameter: 900 mm.
- Pile lengths: 7–9 m.
- All piles equipped with CSL and GGL tubes.

Bryan, Texas, USA (2), FHwA study, 1989–1991 (Baker *et al.*, 1993)

- Stiff clay site.

- Soil description: stiff to very stiff clay with silt and sand seams; average undrained shear strength increasing from 0.8 to 1.5 kN/m^2 (with depth).
- Number of piles: 4 cast-in-place drilled shafts; water/bentonite with tremie concrete placement.
- Pile diameter: 900 mm.
- Pile lengths: 7–9 m.
- All piles equipped with CSL and GGL tubes.

The third (FHwA) set of studies has left a lasting impression on the quality control of new drilled shaft construction. As a result of the conclusions drawn from the different NDT methods applied by various commercial testing houses to these shafts, recommendations were made for nondestructive testing specifications, which have been adopted by several civil authorities in the USA. The four principal small-strain NDT methods tested in these programs were Sonic Echo, Sonic Mobility, Cross-Hole Sonic Logging and single-tube Gamma–Gamma Logging (California only).

The main conclusions drawn from these studies were as follows:

- Whatever the test employed, to extract the most from the nondestructive test programs use should be made of all available information such as shaft length, concrete mix plus construction procedure, plus site records (including records of any problems encountered during construction, as well as theoretical versus actual concrete volume measurements).
- All of the NDT methods employed required well-trained individuals with experience in interpretation of shaft response signals and are thus operator-dependant to some degree (see Chapter 12 on 'Reliability' in this book).
- Major defects and reductions in cross-section of 50 % or more were detected by all of the methods employed.
- Smaller defects and reduction in cross-section were not always detected by some of the techniques.
- The success rate of finding defects depended on the technique. CSL and Gamma–Gamma Logging were more effective in detecting incompletely concreted shaft bottoms and multiple defects with depth, including relatively small defects down to 12 % of shaft cross-sectional area.
- On the other hand, these down-hole methods cannot detect bulbs or defects outside the shaft 'rebar' cage, whereas the Sonic-Echo and Sonic-Mobility tests can, if these anomalies are of sufficient size.
- Sonic-Echo and Sonic-Mobility tests are limited on shafts with high length/diameter ratios.

This led to the conclusion that Sonic Logging or Gamma–Gamma Logging in preplaced access tubes, combined with careful observation of quality control, is necessary and is most cost-effective on highly stressed shafts where small defects could lead to shaft overstress. For larger-diameter shafts, the efficiencies of these two methods could be increased by increasing the number of access tubes. This conclusion led to recommendations that integrity testing is necessary or desirable when one or more of the following conditions are present:

- A very 'high-design' stress level so that only very minor flaws can be tolerated.
- Complex and unpredictably variable soil and water conditions.
- Relative inexperience of the contractor and inspector with the construction techniques being followed.

These recommendations are the framework for most NDT integrity specifications used in the USA at the present time.

Three Class-A studies were supported in the last decade of the 20th Century in several University Research Centers, i.e. in Delft, Holland, Houston, Texas, USA and Amherst, Massachusetts, USA, respectively. The immediate conclusions drawn by the researchers in these three projects do not appear to have influenced the deep foundation industry to the same extent as the FHwA study.

A rather bizarre competition to study the ability of stress-wave methods for determining the integrity of piles was organized by Delft University, The Netherlands (Wheeler, 1992). The pre-cast piles were installed in a soft-soil site with hidden defects, and then the testing organizations were asked to submit their systems to evaluate the piles. However, in order to eliminate any 'personal factor effect' from the test operator, each system was operated by a 'lay person' with no previous knowledge of the technique! The whole operation was completed in a tent covering the piles, and so no 'contamination' of the procedure could be introduced.

Needless to say, there was considerable controversy as to the value of this competition, as witnessed by the discussions in the journal, *Ground Engineering* (Stain, 1993; Turner, 1993; van Weele, 1993). The results of this particular exercise do not appear to have influenced the pile testing community in any way.

In 1996, as part of a study sponsored by the International Association of Foundation Drilling and the FHwA on how minor anomalies in drilled shafts affect the strength of the shafts, the University of Houston, Texas, constructed 22 600-mm-diameter shafts, with various defects at different locations simulated with 25-mm-thick rubber inserts. Several teams from testing firms and universities were invited to participate in the study, with the principal methods employed being Sonic Echo and Sonic Mobility.

Samman and O'Neill (1997a,b) concluded that surface techniques can result in a large number of false positives, cannot differentiate between necking and bulging and are highly dependent on operator skill. They further indicated that down-hole techniques could detect defects larger than 15 % of the cross-sectional area of the shaft, if access tubes are installed at the rate of one tube per 300 mm of shaft diameter. This deduction, however, was based on reported data from other Class-A exercises, since none of the piles in the Houston exercise contained access tubes for CSL testing.

Criticism of the shaft and defect construction methods used in the Houston project was made by several of the analysts (Davis, 1997), particularly with respect to the ability of the rubber inserts to reproduce actual defects found in real practice.

The most recent Class-A prediction study was sponsored by the FHwA at the National Geotechnical Experimentation Site at the University of Massachusetts in Amherst. The results were reported by Iskander *et al.* (2001). Six 8.2-m-long drilled shafts were installed by rotary auger, with diameters varying between 900 mm and 1 m down their lengths. The shafts were reinforced and CSL steel tubes attached to the reinforcing cages in five of the six shafts. The soil profile consisted of 1.5 m of clay fill followed by 2.5 m of sandy silt, in turn overlaying soft to very soft varved clay to below the shaft bases. Different defect types, such as necking, voids and soft bases, were built into the shafts using a variety of materials, including plastic pails and tubes, wool insulation, cardboard and bags filled with soil.

Six teams from commercial testing firms and universities participated in the study, with a mixture of surface and down-hole methods. The performances of the two main method types were evaluated separately.

CSL testers were able to locate and size defects within the shafts exceeding 10 % of the shaft cross-sectional area. Defects smaller than 5 % were typically not detected by the CSL test. The method was able to locate necks extending inside the reinforcing cage, but obviously necks outside the cage were not detected, since they did not extend into the direct path between any pair of access tubes. Soil inclusions were more difficult to detect than voids. Soft bases were detected when access tubes passed through them. Some dependence on operator skill was noted, with reporting of poor-quality concrete where no defects were planned. The fact that different testers reported poor-quality concrete at different locations suggests that these reports were 'false positives'.

Iskander concludes that surface techniques performed better than reported in earlier studies (Baker *et al.*, 1993; Samman and O'Neill, 1997a,b). Surface techniques were able to identify several defects in each shaft, where some defects were as small as 6 % of the shaft cross-sectional area. However, the methods are highly dependent on the skill of the operator, and improved performance in this study may be at least partly attributed to the skill and experience gained by the operators since the earlier Class-A predictions, and improvements made to both the equipment and analysis algorithms. Up to three defects per shaft situated under one another were located; however, up to six defects per shaft were present and no participant located more than three. Small voids were relatively easy to locate, as opposed to larger soil inclusions. The latter were situated deeper in the shafts than the former.

Only one participant analyzed mobility data using the Impedance-Log method (Davis, personal communication, 2000), and these results were not reported by Iskander *et al.* (2001). However, the Impedance-Log analysis was able to locate soft bases where installed, whereas standard surface tests generally did not find them.

Perhaps the major lesson to be learnt from these Class-A prediction exercises over the last thirty years is that, when used to quantify anomalous shaft conditions, the general engineering community is still not convinced about the accuracy and 'independence' of these techniques. Perhaps the most difficult aspect of these methods is the elimination of the 'personal factor' in test-result interpretation. It is the present authors' opinion that this 'personal factor' will always be present, but can be reduced to a minimum by judicious choice of test methods and by training and pre-qualification of testing groups (see Chapter 14).

The series of Class-A prediction exercises is not over yet! At the time of writing this manual, German engineers have recently completed a site at Horstwalde near Berlin recreating piled foundations beneath existing structures, to 'facilitate the development and improvement of assessment methods for piles and enhance their reliability' (Niederleithinger and Taffe, 2003). This site was partly funded by the European Union project RUFUS ('Re-use of Old Foundations on Urban Sites'). The number, location and types of defect in the piles constructed have not yet been announced. It is to be hoped that the problems experienced in designing previous exercises were considered, particularly the difficulty in creating realistic defects in shafts.

12

The Reliability of Pile Shaft Integrity Testing

Thorburn and Thorburn (1977) reviewed the possible causes for observed defects in drilled shafts, and there has been considerable discussion about reliability of the application of nondestructive testing to detecting the presence and engineering significance of shaft defects, and hence the quality assurance of deep foundations (Preiss and Shapiro, 1981; Fleming, 1987; Williams and Stain, 1987; Starke and James, 1988; Baker *et al.*, 1993; Turner, 1997; Cameron and Chapman, 2004).

'Reliability' has a specific meaning when applied to the study of engineering systems by statistical methods. The reliability of a system is the level or degree of confidence that the system will perform as it was designed or intended to do. This can be expressed mathematically (statistically) by considering the reliability of individual system components, together with their relative effects on the performance of the total system.

Nondestructive testing programs can be designed to increase the confidence (and hence, reliability) in the foundation system under construction. Cameron *et al.* (2002) critically examine the statistical sampling methods available to help the engineer in planning NDT programs, and make recommendations for the most suitable statistical approach to be applied. Although their work refers specifically to shaft head impact tests, a statistical approach applies equally to cross-hole and down-hole test methods. They also support Baker *et al.* (1993) in distinguishing between two families of drilled shafts:

- Multiple shafts in groups used to support individual loads.
- Single shafts where the occurrence of defects is more critical for structural performance.

Nondestructive Testing of Deep Foundations B.H. Hertlein and A.G. Davis

Nondestructive testing (NDT) when used in quality assurance testing of piles and drilled shafts can be viewed from two extreme viewpoints. Either NDT helps *Engineers* to confirm that their design and performance criteria are being met, or NDT checks that the *Contractor* has supplied the *Owner* with the product that he has paid for (material quality, minimal geometric requirements, etc.).

From the first point of view, certain deviations from construction specifications can be tolerated and compensated for, provided that there is strong interaction between *Engineer*, *Contractor* and *Tester*. In the second case, no margin for error outside the contract specifications can be allowed; they are 'set in stone'.

Davis (1998) contends that the NDT methods now available are more suited to the first scenario (Performance Specification). This is because the variables inherent in drawing conclusions from NDT on piles are numerous, and it is not always possible to clearly define these variables to reach a complete answer to satisfy the materials specification. The two approaches of Cameron *et al.* (2002) and Davis (1999) are complementary and should be considered together. Both approaches are described here.

12.1 STATISTICAL NDT SAMPLING SCHEMES[1]

The options available when planning an NDT survey are to examine all or none of the piles, or a selected sample. Cameron *et al.* (2002) discuss the merits and disadvantages of these three possibilities, and their statistical considerations are described here.

Fundamental sampling theory includes how the information contained in a statistical sample can be used to draw conclusions about the population from which the sample was taken. In the case of piling NDT, the objective is to determine the percentage of shafts needed for testing and to achieve a pre-defined level of confidence in the construction quality of all the shafts in the population. The NDT test program is assumed to consist of a series of independent, repeated tests, whose results indicate if a shaft is sound or defective. Such quality characteristics are known as 'attributes'. Since only two outcomes of the test are possible, representing the behavior of a discrete random variable, and the probability of detecting a defective shaft is assumed to be constant throughout, then the testing procedure satisfies the requirements of a Bernoulli sequence. The sampling process can be represented mathematically by the hypergeometric distribution. This distribution is the basic model used in conjunction with quality control (Ang and Tang, 1975).

The hypergeometric distribution in equation [12.1] returns the probability of detecting exactly n_d defective shafts in a tested sample of n shafts, given that there are N_d defective shafts in a group of N shafts:

$$P(X = n_d, n/N_d, N) = \frac{\binom{N_d}{n_d}\binom{N-N_d}{n-n_d}}{\binom{N}{n}} \qquad (12.1)$$

[1] After Cameron *et al.*, 2003.

where N is the number of shafts in a selected group, N_d the number of defective shafts in the group, $(N - N_d)$ the number of sound shafts in the group, n the number of shafts in the tested sample, n_d the number of defective shafts in the tested sample, $(n - n_d)$ the number of sound piles in the tested sample and $P(C) = 0.1$ (confidence level, $90\,\% = 1 - P(C)$).

The value of $P(C)$ is equal to the probability of having exactly N_d defective shafts in the entire group of N shafts, given that exactly n_d defective shafts are detected in the tested sample of n shafts, as shown in equation (12.2):

$$P(X = N_d, N/n_d, n) = P(C) \tag{12.2}$$

Therefore, the probability of having more than N_d defective shafts in the entire group of N shafts, under the same conditions, must be less than $P(C)$, as shown in equation (12.3):

$$P(X > N_d, N/n_d, n) < P(C) \tag{12.3}$$

Accordingly, for a sample size n, containing n_d defectives, the probability of having N_d or more defectives in the entire group is less than or equal to (no greater than) $P(C)$, as shown in equation (12.4):

$$P(X \geq N_d, N/n_d, n) \leq P(C) \tag{12.4}$$

The size of the group (N), number of defective shafts that can be tolerated within the group (N_d) and the number of defectives detected during testing (n_d), must be specified beforehand, in order to perform the analysis.

Figure 12.1 shows percentage NDT plotted against defectives in the group, for various numbers of defectives identified during testing within a group of one hundred

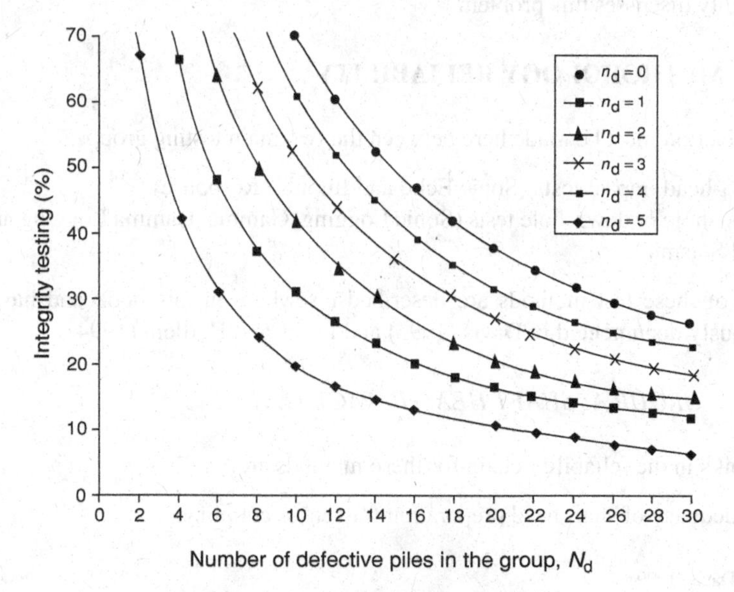

Figure 12.1 Level of confidence analysis

shafts to achieve a confidence level of 90 %. As an example, by testing 20 % of the shaft group and identifying no defective shafts, the likelihood of there being more than ten defectives in the entire group is less than 10 % (Preiss and Shapiro, 1979). In other words, we are 90 % sure that there will be less than ten defective shafts based on the sampling results. By testing 60 % of the group and identifying no defectives, the likelihood of having more than three defectives in the entire group is less than 10 %. The level of confidence in the quality of the group grows as the amount of testing increases. However, if the amount of testing is reduced, one must accept a higher probability of there being more than N_d defectives in the group.

In order to perform this analysis, it is assumed that the shaft population is divided into groups prior to testing. This process of stratified sampling ensures that shafts will be selected from all locations over a particular site. The sample to be tested is assumed to be selected completely at random from within a group; however, in practice there are a number of factors that may influence this decision. These factors include areas of particularly poor ground on a site, when there are known problems during construction and where shafts are at locations where differential settlement is critical. It is also assumed that the likelihood of constructing a defective shaft is the same (uniform) throughout a given site, irrespective of the varying ground conditions and other uncertainties involved.

Taking into account not only the probability of constructing a defective pile, but also the probability of accurately detecting a defective pile using low-strain integrity testing, could make improvements to this statistical sampling approach. Implicit in this approach is the assumption that the integrity testing procedure is totally reliable and that the test results are absolutely accurate. The following review of test methodology reliability discusses this problem.

12.2 METHODOLOGY RELIABILITY[2]

A distinction must be made here between the two main testing groups:

- Shaft-head impact tests (Sonic Echo and Impulse Response).
- Cross-hole or down-hole tests (Sonic Logging, Gamma–Gamma Logging and Parallel Seismic).

All of these test methods are described elsewhere in this book and have been previously documented in Davis (1995) and Davis and Hertlein (1994).

12.2.1 GROUP A: SHAFT HEAD IMPACT TESTS

The links in the reliability chain for these methods are:

(1) Adequacy of shaft head preparation for sensor coupling.

[2] After Davis, 1999.

(2) Correct and recent calibration of load-cells and sensors.
(3) Suitable data-acquisition, signal-filtering and processing systems.
(4) Trained and experienced site operators.
(5) Degree of signal damping (high shaft length/diameter (l/d) ratios, stiff lateral soils and bulges in the upper portion of the shaft all cause high signal damping).
(6) Presence or not of multiple anomalies down the shaft length (an anomaly can be a bulge, a neck-in, cracking, honeycombing or soil/laitance inclusions).
(7) The experience of the testing engineer in test data interpretation (the 'personal' component).

These factors can be combined in a 'lumped' reliability model, and efforts are now being made to adapt such models to this problem. Any breakdown of a single component in this model can be seen to throw the system reliability into question.

Links (1) through (4) can be addressed and improved by the Construction Industry (DFI, ADSC, FHwA, ACI) by providing training and certification programs for method, operator and testing company. This would help to ensure that correct and recognized procedures are observed at all times.

Problems associated with links (5) through (7) can be minimized by test-method improvement, both in hardware and in software. Examples include the possible introduction of 'lost' sensors at positions down selected shafts during construction (hardware) and the introduction of improved data analysis by methods such as the Impedance Log (Paquet, 1991).

The present state-of-the-art for Group-A test methods suggests that they are very reliable when shafts have l/d ratios <30/1 in relatively soft soils, and when any significant defects are limited to the upper two thirds of the shaft length. This reliability drops rapidly when stiff lateral soils are combined with high l/d ratios, and defects are either in the bottom third of the shaft, or are multiple.

12.2.1.1 Signal-to-Noise Ratio

The response sensor signal (either geophone or accelerometer) is the most important parameter affecting the reliability of the method. This is the signal that carries information about any anomalies that are present in the pile shaft. The quality of this carrier signal is a direct function of the signal-to-noise ratio, which is influenced by items 1, 3, 4, 5 and 6 in Group A above.

Item 1 – Adequacy of Shaft Head Preparation

Inadequate shaft head preparation (lack of smoothness, concrete 'micro-cracking', etc.) results in an increase in the amount of noise generated (i.e. a decrease in the signal-to-noise ratio) and can mask the responses from deeper anomalies altogether. The example trace in Figure 12.2 is from a velocity transducer and shows a predominant, logarithmically decaying voltage–time signal typical of near-surface generated noise, usually from cracked concrete.

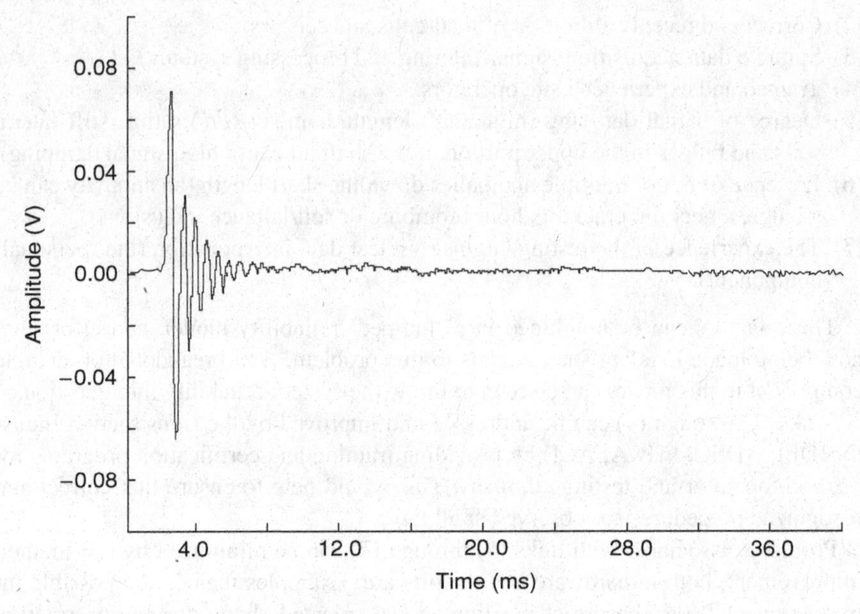

Figure 12.2 Voltage–time signal trace, typical of near-surface generated noise

Item 3 – Suitable Data Acquisition, Signal Filtering and Processing Systems

Filtering of the response signal in the acquisition unit can either be analog ('up-front') or digital, or a combination of both. If the filtering is not sufficient, too much noise remains on the signal trace. If the filtering (smoothing) is too severe, important information from deeper anomalies can be removed. If analog filtering is employed, the original signal cannot be recovered at a later date. It is important to conserve as much of the 'raw' signal trace as possible, for future signal processing by different methods. Figure 12.3 represents an example where excessive filtering of the lower time signal producing the upper trace has resulted in loss of significant information. On the other hand, Figure 12.4 shows the benefits of optimal signal filtering.

Item 4 – Trained and Experienced Site Operators

The operator controls the positioning of the sensor on the pile head and can observe whether the concrete in the pile head is in good shape for satisfactory signal reception.

Item 5 – Degree of Signal Damping

The geometrical characteristics of the pile control the degree of signal damping (reduction in signal-to-noise ratio). The greater the *l*/*d* ratio and the stiffer the

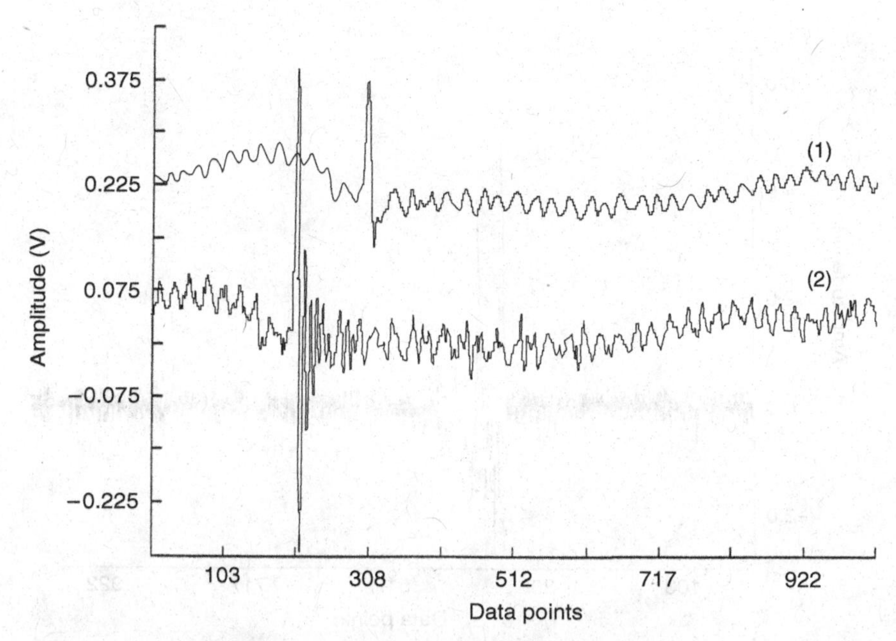

Figure 12.3 Voltage–time signal traces, where excessive filtering of raw signal (2) loses information (1)

lateral soils, the lower the signal-to-noise ratio from anomalies at depth. Eventually, the response signal becomes so damped that no information below a certain pile depth is attainable. Examples of this extreme damping are given in Figures 12.5(a) and 12.5(b) for both Sonic-Echo and Impulse-Response testing, respectively. This damping removes any trace of resonant frequency peaks in the mobility plot in Figure 12.5(b).

Item 6 – Multiple Anomalies

The signal from the first anomaly encountered down a pile shaft is much stronger than the signal from a deeper anomaly, and can mask the second response completely (see Figure 12.6). In mobility–frequency plots, the second, weaker response often 'rides' on the first signal as small-amplitude peaks superimposed on the more dominant peaks (see Figure 12.7). There are techniques for enhancing the second signal, but these are often not good enough to produce a satisfactory signal/noise ratio.

12.2.1.2 Personal Factor

The personal component applies both to the test operator and to the engineer responsible for test data interpretation. These can be the same individual.

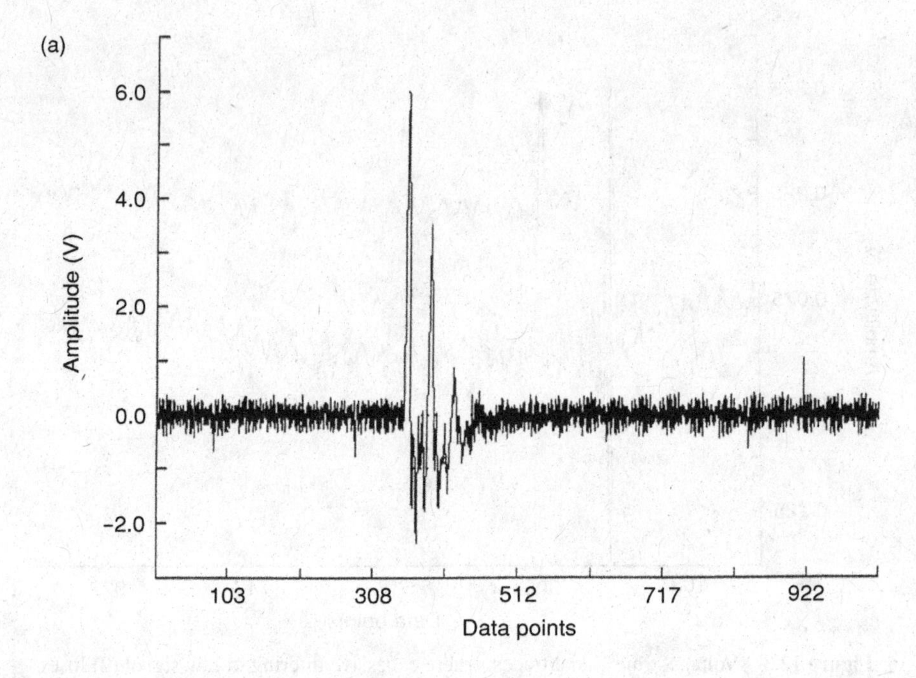

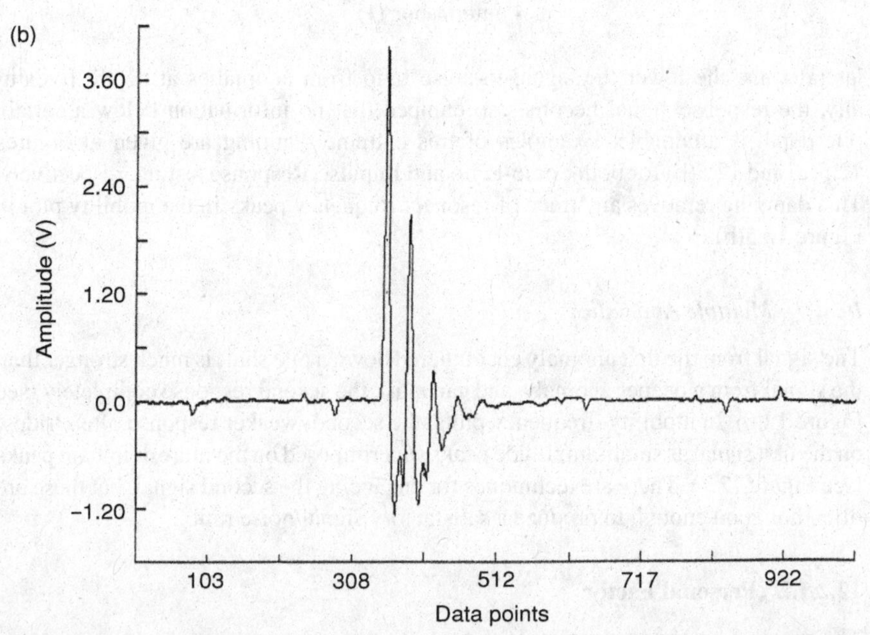

Figure 12.4 Voltage–time signal traces: (a) unfiltered raw signal; (b) trace from (b) with optimal filtering applied

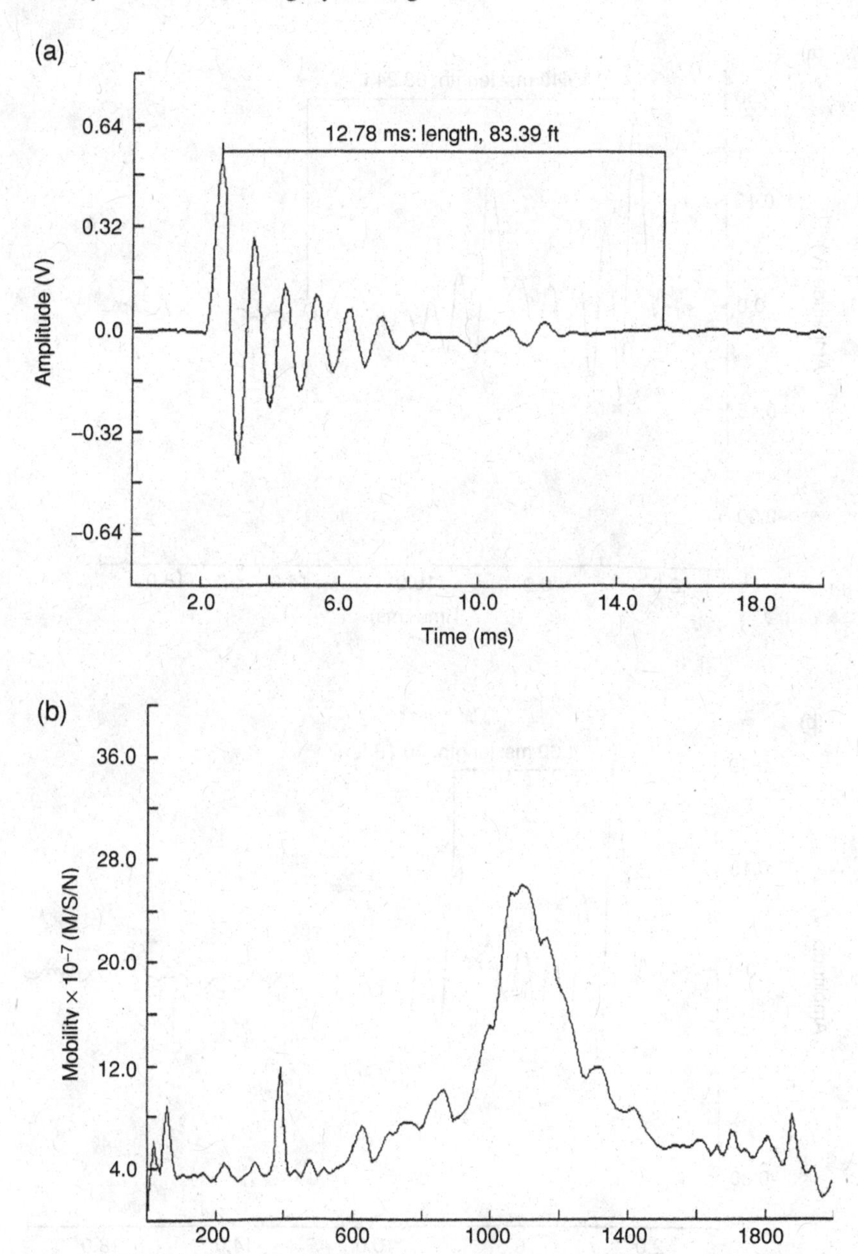

Figure 12.5 Effect of extreme damping on (a) Sonic-Echo and (b) Impulse-Response data

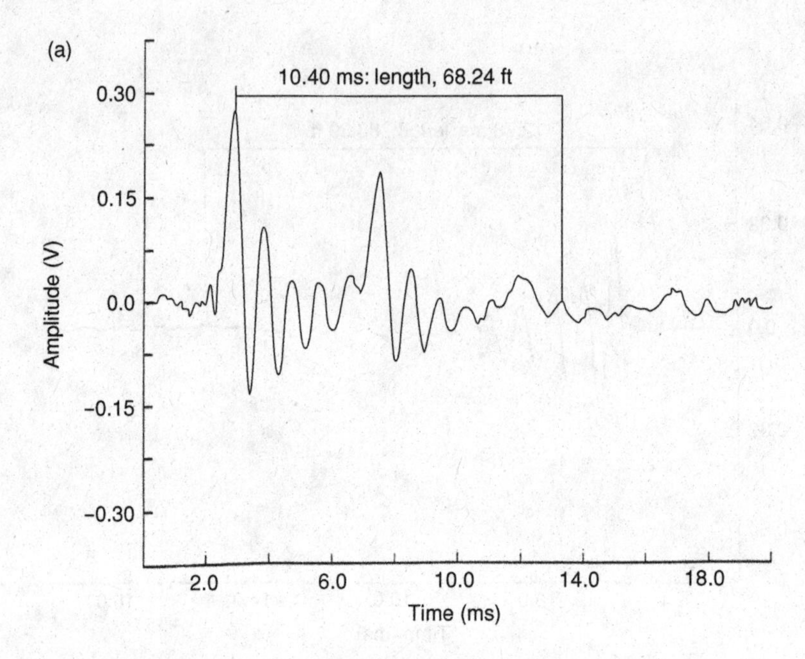

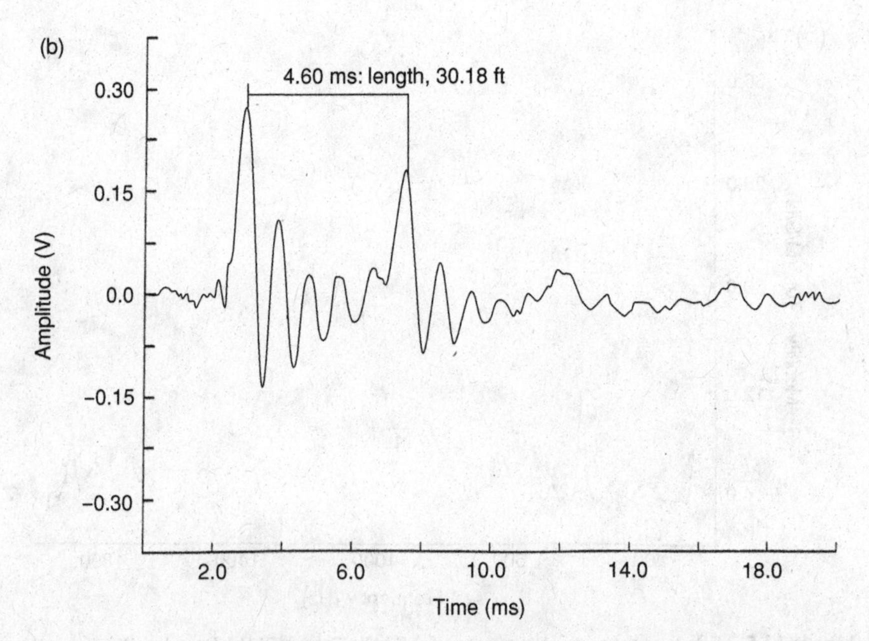

Figure 12.6 (a) Secondary reflection masking toe reflection and (b) depth of secondary reflector in Sonic-Echo data

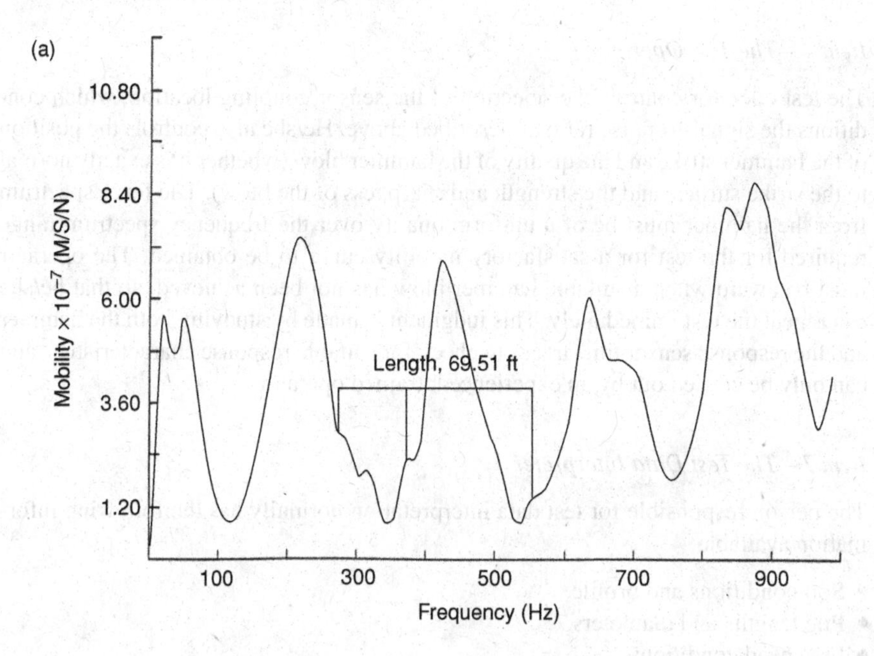

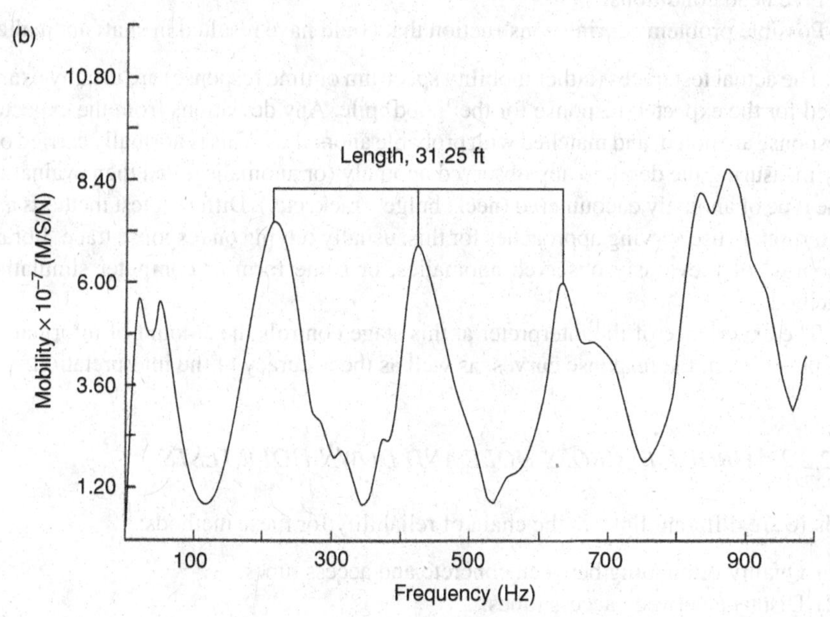

Figure 12.7 (a) Secondary reflection masking toe reflection and (b) depth of secondary reflector in Impulse-Response data

Item 4 – The Test Operator

The test operator controls the selection of the sensor coupling location, which conditions the signal-to-noise ratio as described above. He/she also controls the position of the hammer strike and the quality of the hammer blow (whether it is exactly normal to the strike surface and the strength and sharpness of the blow). The force spectrum from the hammer must be of a uniform quality over the frequency spectrum range required for the test for a satisfactory mobility curve to be obtained. The operator must be aware when a suitable hammer blow has not been achieved, so that he/she can repeat the test immediately. This judgment is made by studying both the hammer and the response sensor time traces to check for suitable response characteristics, and can only be carried out by an experienced, trained operator.

Item 7 – The Test Data Interpreter

The person responsible for test data interpretation normally has the following information available:

- Soil conditions and profile.
- Pile lengths and diameters.
- Free head conditions.
- Possible problems during construction that could have resulted in shaft anomalies.

The actual test traces (either mobility spectrum or time response) are usually examined for the expected response for the 'good' pile. Any deviations from the expected response are noted, and matched with probable anomalies. This is normally carried out by measuring the depth to any observed anomaly (or anomalies) and then evaluating the type of anomaly encountered (neck, bulge, crack, etc.). Different test methods and interpreters use varying approaches for this, usually relying on response trace 'library records' of previously observed anomalies, or some form of computer simulation method.

The experience of the interpreter at this stage controls the amount of information obtained from the response curves, as well as the accuracy of the interpretation.

12.2.2 GROUP B: CROSS-HOLE AND DOWN-HOLE TESTS

There are different links in the chain of reliability for these methods:

(1) Quality of bonding between concrete and access tubes.
(2) Distance between access tubes.
(3) Number of tubes per unit area of shaft cross-section.
(4) Distance of perimeter access tubes from shaft perimeter.
(5) Experience of the testing engineer in the interpretation of test data (personal factor).

As for Group A, it is possible to define 'lumped' reliability models for these tests. Links 1 through 4 can be addressed by careful pre-planning for the material used for the tubes, as well as their number and location.

12.2.2.1 Signal-to-Noise Ratio

Items 1 and 2 in the introduction to Group B above affect this.

Item 1 – Bonding of Access Tubes to Concrete

The use of plastic PVC tubes in drilled shafts for both Gamma–Gamma Logging and Sonic Logging has often resulted in poor bonding between the concrete and the PVC under certain concrete-curing conditions, particularly over the top ten to twenty feet of the shaft. For Gamma–Gamma Logging, this results in some loss of signal penetration into the surrounding concrete. In the case of Sonic-Logging and Parallel-Seismic tests, there is usually a total loss of signal, because of the air gap formed around the tubes. An example of loss of signal as a result of tube debonding is given in Figure 12.8.

Item 2 – Distance Between Access Tubes

This is of particular importance for Cross-hole Sonic Logging, where damping of the signal occurs with increasing tube spacing. Critical tube spacing is reached where the

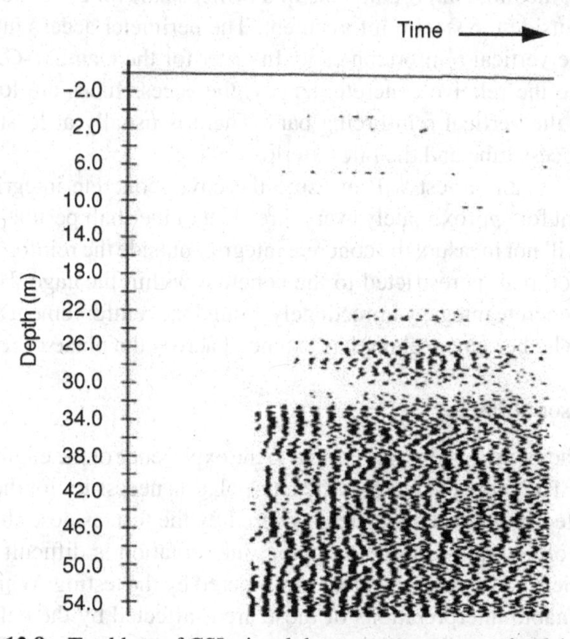

Figure 12.8 Total loss of CSL signal due to severe access tube debonding

signal is totally damped by the surrounding concrete. For most available systems, this is usually at distances of 3.6 m (12 ft) or greater.

12.2.2.2 Area of Shaft Tested

This concerns Items 3 and 4 in Group B above.

Item 3 – Number of Tubes per Unit Area of Shaft Cross-Section

This affects the shaft area covered by both Sonic-Logging and Gamma–Gamma Logging tests.

The Gamma–Gamma test irradiates a concrete cylinder within a radius of approximately 90 mm (3.5 in) around the tube. Normal tube density is one tube for each 300 mm (1 ft)-diameter of pile shaft, i.e. for a 1.8 m (6 ft)-diameter shaft, there will be six tubes. Therefore, for a shaft with a 2.8 m^2 (28.25 ft^2) cross-sectional area, only 0.16 m^2 (1.6 ft^2) will be tested in normal practice (less than 6 % of the total concrete volume).

For the same tube density, the Cross-hole Sonic-Logging technique tests approximately 1.3 m^2 (12.8 ft^2) when all of the perimeter and diagonal paths are tested. This is still only 45 % of the total concrete volume in the shaft.

Item 4 – Distance of Perimeter Access Tubes from Shaft Perimeter

In many cases, the most important zone of a drilled shaft for concrete integrity is the 'covercrete' outside the steel reinforcement. The perimeter access tubes are usually fixed inside the vertical reinforcing cage. In order for the Gamma–Gamma Logging test to measure the relative concrete density, the access tubes are located as far as possible from the vertical reinforcing bars. There is usually at least 75 mm (3 in) between the access tube and the pile exterior.

The Gamma–Gamma test will measure the cover concrete integrity at each test point (one point for approximately every three feet of the shaft perimeter). The Sonic-Logging test will not measure the concrete integrity outside the reinforcing cage, since the transmission path is restricted to the concrete within the cage. However, it will ascertain the concrete integrity immediately around the reinforcement and will be able to determine whether any necking has extended across the plane of reinforcement.

12.2.2.3 Personal Factor

The personal factor in these tests is limited to the experience of the engineer in test-data interpretation. The principal factor is the interpolation necessary for the interpretation of concrete integrity between the areas covered by the test, as described above.

In the case of Gamma–Gamma Logging, interpolation is difficult without added help from vertical coring in those areas not covered by the testing. While it is possible to reach reasonable interpretations of those areas affected by the defects located in Sonic Logging, it is usually still necessary to confirm the type of defect by selected vertical coring.

13

Current Research

Most of the fundamental research into and development of the nondestructive integrity test methods in use at the present time was undertaken in the two decades between 1965 and 1985. National research organizations such as CEBTP in France and TNO in Holland funded extensive research and development programs, supported by large geotechnical or deep foundation companies such as Fugro in Holland, Solétanche and Bachy in France, Cementation Piling in the UK and GRL and Associates in the USA. Some research activity was evident at Universities such as Edinburgh University in Scotland, UK and the University of Texas at Austin, in the United States, at the instigation of professors with specific interests in this field. However, most progress in developing and implementing the methods for commercial application was made using funds from outside the academic arena.

The sources for this type of research funding mostly dried up after 1985 and significant research programs disappeared. Research activity has continued at a reduced level, principally in testing companies that have allocated portions of their budgets to research and development, and in a few academic establishments, mainly in the United States. A great deal of national funding for nondestructive testing research over the last fifteen years has been dedicated to Class-A prediction studies on shafts with built-in defects (see Chapter 11 – 'Class-A Predictions'), as opposed to fundamental research into new methods. Commercial research over the last decade or so has concentrated on equipment miniaturization and reliability, as well as speeding-up data acquisition and analysis for the currently approved methods. Fortunately, at least one group managed to find funding for research aimed at advancing the application of the low-strain methods.

Nondestructive Testing of Deep Foundations B.H. Hertlein and A.G. Davis

13.1 DEVELOPMENTS IN MEASUREMENT AND ANALYSIS

13.1.1 THE IMPORTANCE OF TRANSDUCER COUPLING

Recent research in Germany has shown that German engineers are also taking deep foundation quality control seriously. A paper presented at the 2002 ASCE Geotechnical Conference in Orlando, Florida, USA described some important research into the minimum sizes of anomalies that could be detected by the current generation of low-strain tests (Kirsch and Plassman, 2002). When developing the Impulse-Response test in the 1970s, engineers at the CEBTP in France realized that the hardness of the material from which the impact hammer head was made determined the duration of the impact and the frequency range of the energy generated. The CEBTP also determined that the rigidity of the velocity transducer attachment affected the useful frequency range of the recorded data. Neither attribute was analyzed in much detail, but the authors of this present book, through many site applications, have consistently proven the value of the extended frequency range afforded by a suitably rigid attachment of the geophone, and selection of an appropriate hammer tip material for the information being sought. Kirsch and Plassman, funded by The German Research Society (DFG), have quantified both phenomena, incidentally exposing an intrinsic limitation in the way that the Impulse-Echo method is applied by many practitioners.

It is common practice among Sonic-Echo users to simply hold an acceleration transducer (an accelerometer) in contact with the concrete at the top of the shaft by hand while the test is performed. The accepted practice among Impulse-Response test users is to fix a velocity transducer (a geophone that is used in place of the Sonic-Echo method's accelerometer), to sound concrete with either a stiff grease or melted wax. Kirsch and Plassman's work strongly suggests that the wax or stiff grease provide optimum results, whereas the hand-held method results in a much lower useable frequency range which limits the minimum size of anomaly that can be detected, thus reducing the resolution and usefulness of the Sonic-Echo technique.

13.2 ELECTRICAL METHODS

There appears to be a general consensus that physical principles for successful testing are restricted to nuclear and stress-wave methods in their various forms. As a result, other physical systems such as electricity or visual assessment have been largely disregarded. The following summaries describe some tentative steps towards these alternative methods.

In practice, some experimental procedures using electrical methods have shown promise under certain circumstances. It is hoped that this book has made it clear to the reader that every nondestructive test method for deep foundations has it own particular capabilities and limitations. Sometimes a nondestructive or non-invasive test may be the only economically viable means of collecting information that can be used to decide which foundations warrant further, more exhaustive or invasive investigation. A thorough knowledge of physics can help a person recognize when some unconventional but physically sound approach may be able to obtain the required information.

A report on deep foundation testing by A.J. Weltman was published by the United Kingdom Construction Industry Research and Information Association (CIRIA) in 1977 (Weltman, 1977). Weltman identified the four basic techniques for electrical testing of deep foundations by using the reinforcing steel cage as an electrode. In order of increasing complexity they are as follows:

- Self-potential
- Resistance to Earth
- Resistivity (Wenner array)
- Induced Polarization.

All four of these techniques were well-established in the geophysics community as methods of assessing various soil properties and identifying subsurface anomalies or buried structures, using non-polarizing electrodes embedded in the near-surface soil.

13.2.1 NON-POLARIZING ELECTRODES

One of the main keys to success with these techniques is the use of an appropriate type of electrode. The flow of ions through an electrolyte (in this case, the soil) to an electrode results in the deposition of metallic and organic ions on the surface of the electrode if a simple conductor such as a metal rod is used. This ionic deposition will eventually create an electrically resistive layer, reducing or stopping the flow of electricity. This effect is known as 'resistance polarization'. Non-polarizing electrodes are designed to minimize this effect. A non-polarizing electrode consists of a conductive rod immersed in a saturated solution or gel of a salt of the same material. While it is very difficult to manufacture a truly non-polarizing electrode, those that are commercially available have an adequate service life for geophysical survey techniques. Typical non-polarizing electrodes for geophysical work consist of silver in silver chloride (Ag/AgCl) or copper in copper sulfate ($Cu/CuSO_4$). The same basic designs are used to construct 'half-cells' for the detection and measurement of corrosion currents in reinforced concrete.

13.2.2 SELF-POTENTIAL

Conductive materials with dissimilar electrical properties in proximity to one another create a difference in electrical potential that can be measured simply by connecting a sensitive voltmeter in series with them. This physical principle is the driving force behind the most common forms of corrosion of reinforcing steel embedded in concrete.

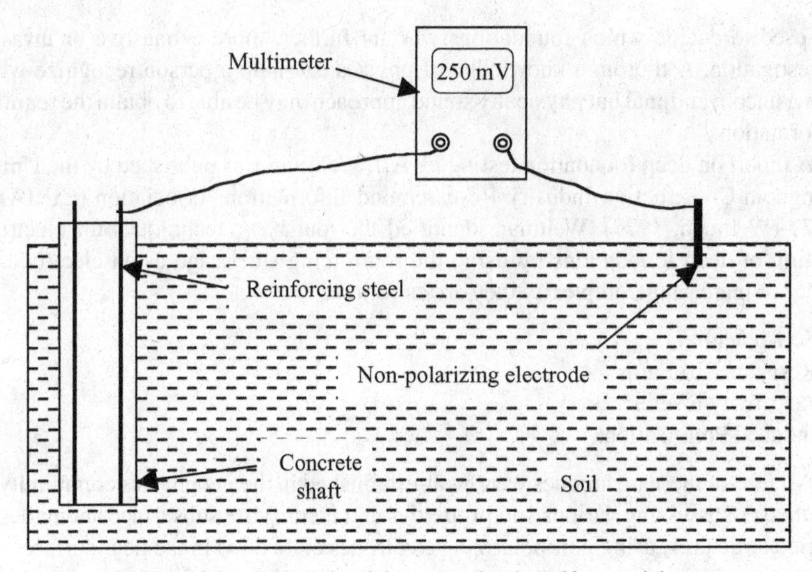

Figure 13.1 Schematic of the set-up for the self-potential test

Since mature concrete is an effective insulator, there should be negligible current flow between reinforcing steel embedded in a concrete shaft and the surrounding soil. If, however, the reinforcing steel is exposed to the soil, either by lack of cover concrete or by moisture infiltration through cracks, and there is enough moisture present in the soil at that point to provide adequate conductivity, a potential difference will be apparent between the reinforcing steel and a non-polarizing electrode embedded in the soil nearby (Figure 13.1).

13.2.3 RESISTANCE TO EARTH

A non-polarizing electrode is embedded in the soil near the foundation and connected to the reinforcing steel via a voltmeter, as for the self-potential test, but an electric current source is added to the circuit to impose a greater potential difference. The higher potential difference can make the test more effective in dryer conditions where the self-potential test is inconclusive (Figure 13.2).

13.2.4 RESISTIVITY (WENNER ARRAY)

The bulk resistivity of the pile/soil structure can be assessed by imposing a current via an electrical source connected between the reinforcing steel and a non-polarizing electrode embedded in the soil nearby. Two additional electrodes are embedded in the soil between the foundation shaft and the current electrode, such that the spacings between all the electrodes, including the pile, are approximately equal. This

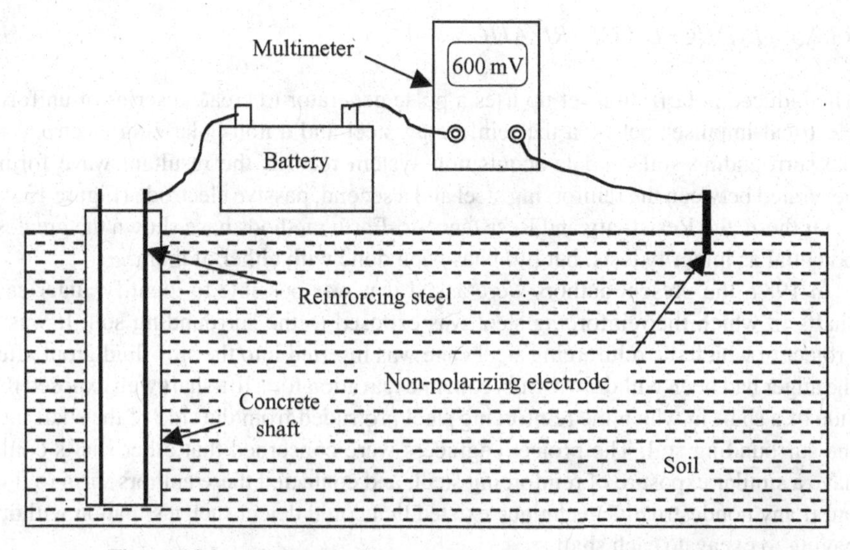

Figure 13.2 Schematic of the set-up for the resistance-to earth test

is known as the 'four-pole Wenner array'. A voltmeter is connected between the intermediate electrodes to measure the potential difference induced by the current flow between the source electrode and the foundation (Figure 13.3). If reinforcing steel is exposed to the soil, higher currents and greater potential differences will be observed.

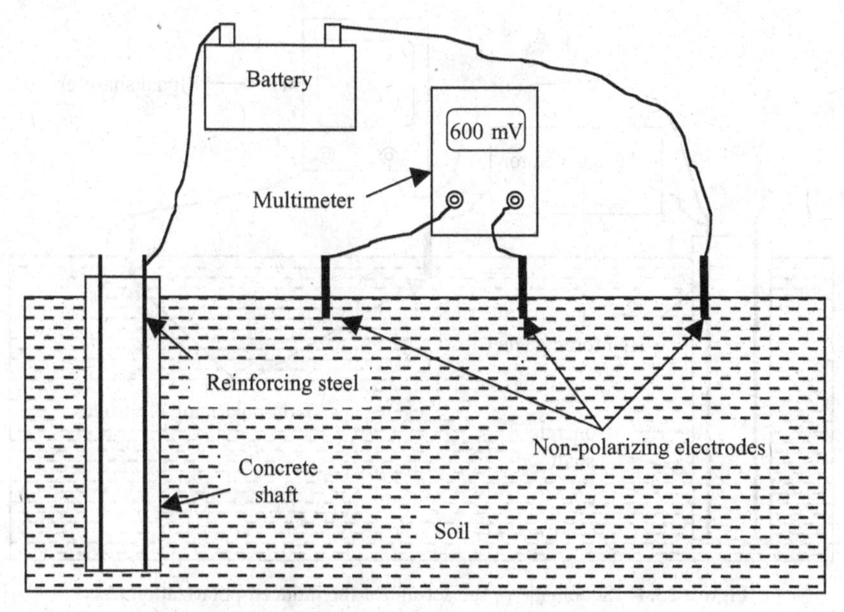

Figure 13.3 Schematic of the set-up for the resistivity test

13.2.5 INDUCED POLARIZATION

The induced polarization set-up uses a pulse generator to create a series of uniform electrical impulses between the reinforcing steel and a non-polarizing electrode in the surrounding soil. A data-acquisition system records the resultant wave forms generated between the reinforcing steel and a second, passive electrode (Figure 13.4).

Of these, the Resistivity and Resistance-to-Earth methods have shown the greatest potential as integrity tests, but little has been done with either of them.

In 1984, the present authors were asked if it was possible to identify auger-cast shafts in which the reinforcing steel was exposed to the surrounding soil. It was a project in which the reinforcing steel cage was inserted into the still-fluid grout after the auger had been withdrawn. Subsequent excavation for a lift-shaft well exposed the side of a shaft in which the reinforcing steel protruded from the side of the shaft into the surrounding soil. The project engineers were concerned that other shafts could have a similar exposure of reinforcing steel, and contacted these authors' firm to find out if any nondestructive technique existed that could detect such a situation without having to excavate each shaft.

The present authors considered the capabilities of existing technology, ruling out all of the then-accepted techniques for integrity testing, since they were incapable of detecting such an anomaly. It was decided to try shaft-to-shaft electrical resistance measurements, reasoning that hardened concrete and grout both have a relatively high resistance compared with moist soil.

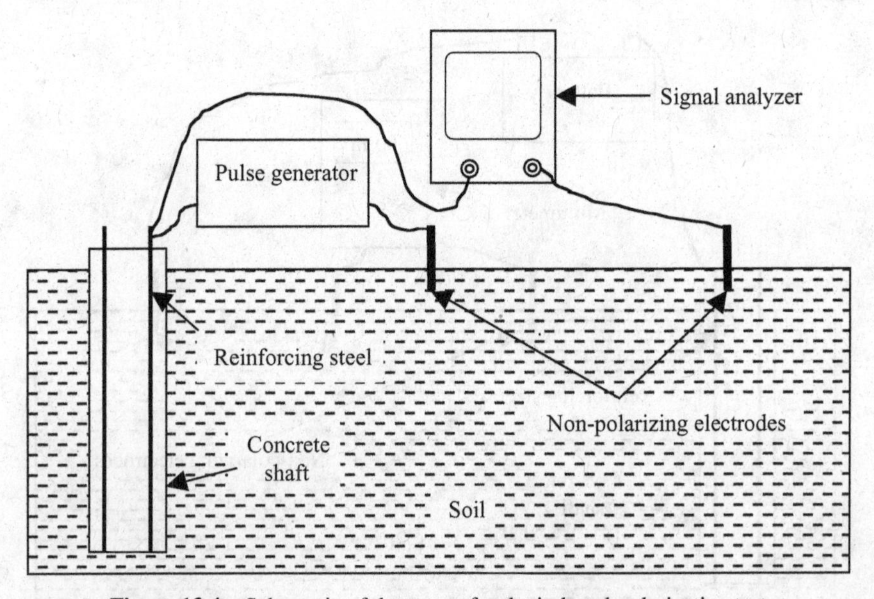

Figure 13.4 Schematic of the set-up for the induced-polarization test

The cover concrete or grout should therefore act as an insulator, or at least a high-value resistor. If the voltage supply is connected across the reinforcing steel in two adjacent piles to form a circuit via the soil between the piles, then two piles with normal grout cover to the reinforcing will have a higher resistance (Figure 13.5(a)), hence lower current draw, than if one or both of the shafts has exposed reinforcing steel in direct contact with the soil (Figure 13.5(b)).

In practice, the technique worked well. Taking a reading between each combination of pile pairs produced a spreadsheet of data that quickly identified exposed steel by a simple process of comparison and elimination. If both shafts were normal, electrical resistance was high, on the order of several kilo-ohms and therefore the current draw was low, on the order of a few hundred milliamps. If the steel was exposed in both shafts, resistivity was very low, typically below 100 ohms, and current draw was in the range 30 to 50 amps. If steel was exposed in only one shaft of a pair, the resistivity value was intermediate, typically several hundred ohms, and current draw was in the range 5 to 10 amps.

13.2.6 CROSS-BOREHOLE RADAR AND ELECTRICAL RESISTIVITY TOMOGRAPHY

The difficulties inherent in determining the integrity of stone columns or deep-mixed soil columns were discussed in Chapter 2 – 'Commonly Used Deep Foundation Construction Methods'. A paper presented by Staab *et al.* at the GeoSupport 2004 Conference in Orlando, FL, USA, however, discussed work done with simulation of Cross-Borehole Ground Penetrating Radar (XBGPR) and Electrical Resistivity (ER) Tomography methods, both of which show some promise for evaluation of soils improved by deep soil mixing (Staab *et al.*, 2004). The computer models generated in the first stage of Staab's research showed that significant defects in a deep-mixed column are likely to be detected by either method, but both were subject to a trade-off between economy of testing and the minimum size of anomaly that could be resolved.

Both test methods require access boreholes to be drilled vertically on each side of the column to be tested, so that the test transducers can be placed at the required depths. For the XBGPR method, two opposing boreholes were used. The transmitting antenna would be in one borehole, while the receiving antenna would be placed in the borehole on the opposite side of the deep-mixed column. The radar impulse thus travels from the transmitter borehole, through the deep-mixed column, to the receiving borehole. To simplify the tomography calculations, the XBGPR data are recorded as a series of discrete measurements, or samples, spaced vertically at 0.25 m intervals (Figure 13.6).

For the ER method, four boreholes were used, one on each side of the column. A string of uniformly spaced electrodes is placed in each borehole. One electrode string is used as the source, and the other three strings are the receivers (Figure 13.7). For Staab's research, the spacing of the ER electrodes was also 0.25 m, the same as the

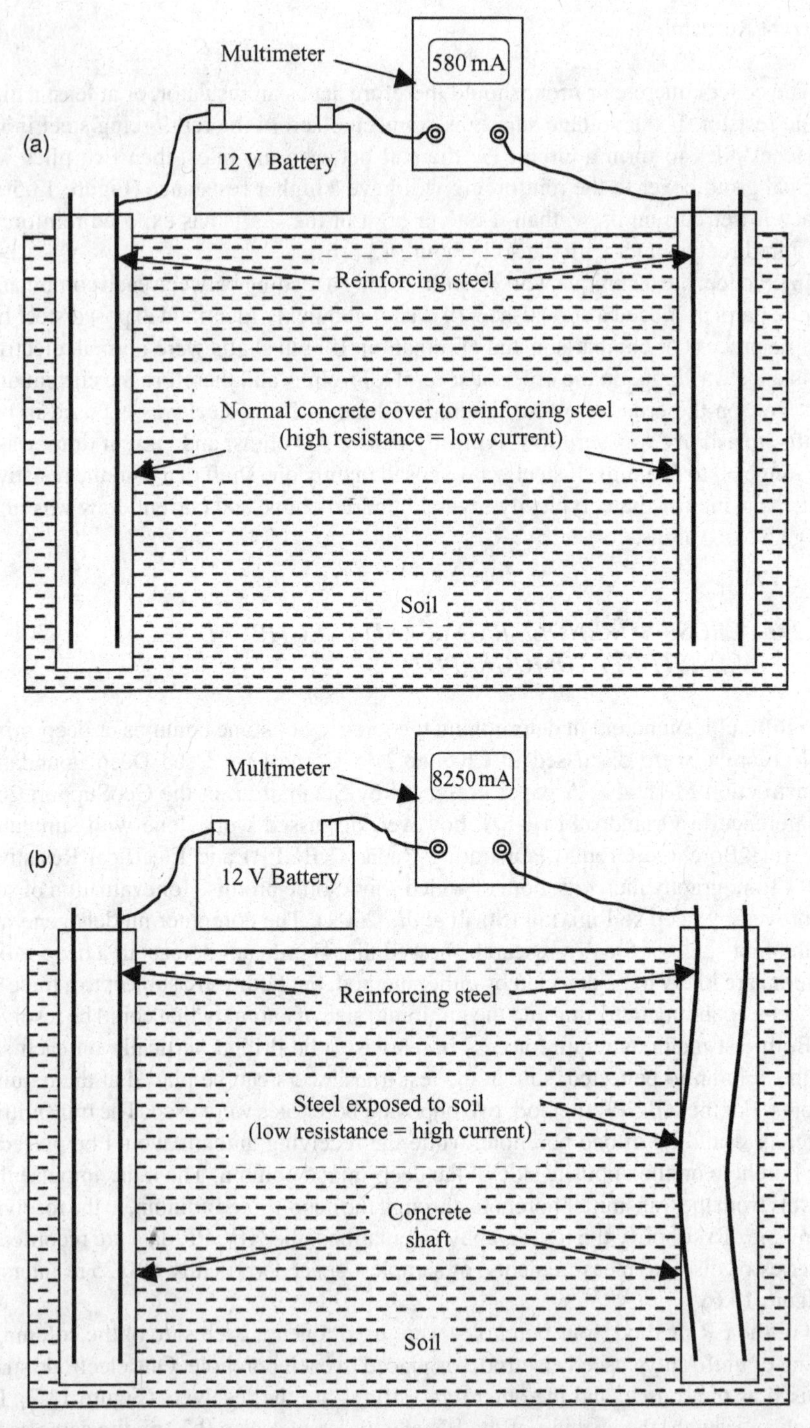

Figure 13.5 Schematics of the set-ups for the shaft-to-shaft resistance text: (a) normal cover concrete; (b) exposed steel

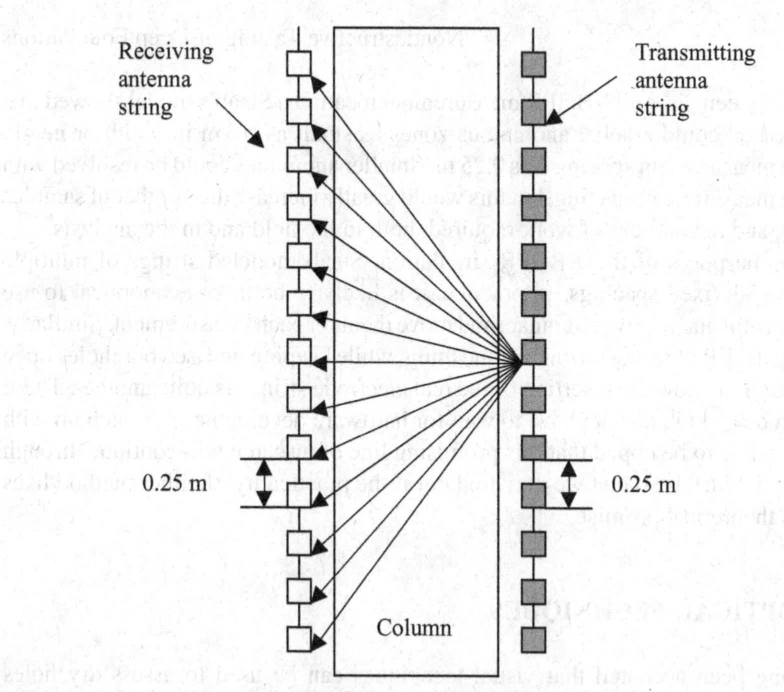

Figure 13.6 Cross-Borehole Ground Penetrating Radar (XBGPR) simulation (after Staab *et al.*, 2004)

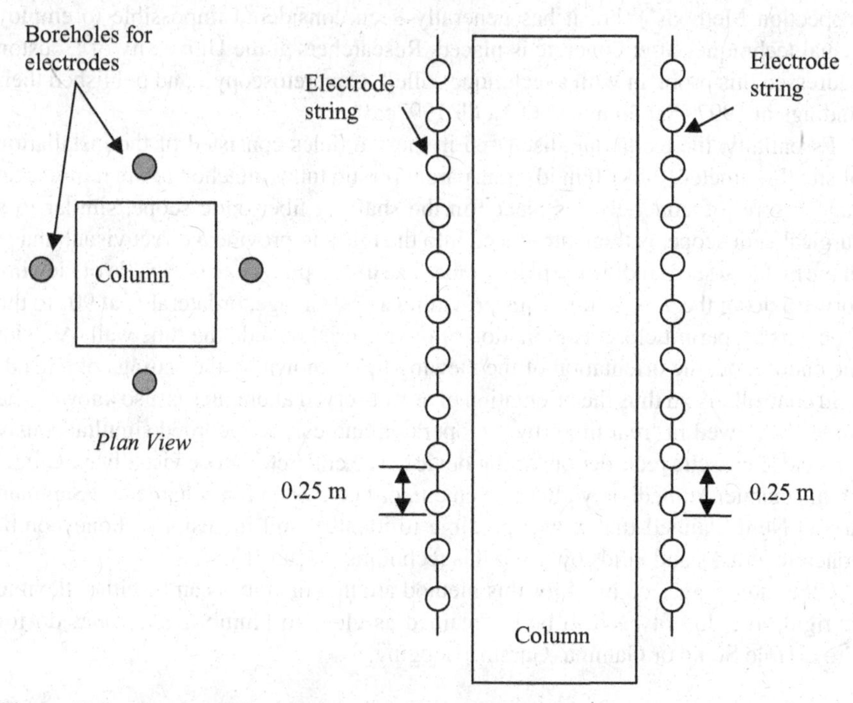

Figure 13.7 Simulated Electrical Resistivity (ER) electrode arrays (after Staab *et al.*, 2004)

spacing between discrete XBGPR measurement locations. Staab's model showed that either method could resolve anomalous zones as small as 0.3 m in width or height when the measurement spacing was 0.25 m. Smaller anomalies could be resolved with a smaller measurement spacing, but this would greatly increase the number of samples recorded, and the amount of work required, both in the field and in the analysis.

For the purposes of the XBGPR simulation, Staab modeled strings of multiple antennae with fixed spacings. In practice, it is likely to be more economical to use single transmit and receive antennae, and move them for each measurement. Similarly, modeling the ER electrode strings is one thing, while keeping uncased boreholes open long enough to allow the insertion of the real electrode strings is quite another. These may both be techniques that have to wait for hardware development to catch up with the theory. It is to be hoped that this promising line of research will continue through full-scale field trials so that we may find out if the practicality of either method lives up to the theoretical promise.

13.3 OPTICAL TECHNIQUES

It has long been accepted that visual techniques can be used to assess dry holes before concrete placement, and in recent years the Shaft Inspection Device (SID) has made inspection of wet shafts possible (see Chapter 4 – 'Traditional and Visual Inspection Methods'), but it has generally been considered impossible to employ visual techniques after concrete is placed. Researchers at the University of Houston addressed this problem with a technique called 'Concretoscopy', and published their findings in 1997 (Samman and O'Neill, 1997c,d).

Essentially, the technique discussed in these articles consisted of the installation of small-diameter (0.5–1.0 in id) transparent plastic tubes attached to the reinforcing cage before the concrete was placed in the shaft. A fiber-optic scope, similar to a surgical endoscope, is then introduced into the tubes to provide a direct visual image of the material surrounding the plastic tube. The fiber-optic scope is capable of viewing forward down the access tube, thus providing a 360° image, or laterally, at 90° to the tube axis, to permit close examination of the material outside the tube wall. As with the endoscope, the orientation of the viewing tip is known by the settings of a hand-held controller, and thus the orientation of any observed anomalies is also known. The image is viewed in 'real time' by the operator, and can be recorded simultaneously on a video cassette recorder or similar device. The concrete can be visually examined in this manner immediately after placement, or later, when it has hardened. Samman and O'Neill claimed that it was possible to identify soil inclusions, 'honeycomb' concrete, cracks and voids by using this technique (Figure 13.8)

Other advantages claimed for this method are that the tubes can be either flexible or rigid, and do not need to be maintained as close to plumb as the tubes do for Cross-Hole Sonic or Gamma–Gamma Logging.

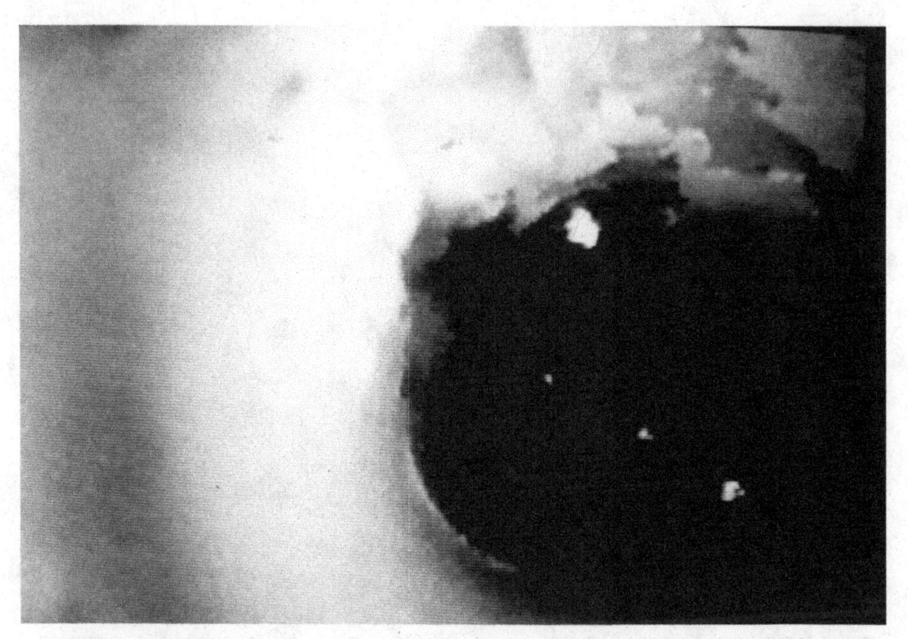

Figure 13.8 Forward view of soil inclusion around the access tube. Reproduced by permission of Stress Engineering, Inc., Texas, USA

In addition to testing drilled shafts, Samman and O'Neill suggested that the method could be used in critical parts of structures to monitor long-term deterioration, such as the development of cracks and/or leaching of concrete by water infiltration (Figures 13.9 and 13.10).

One of the key limitations of the method is that it can provide no information about the material between the access tubes. Only the surface of the material in contact with the access tube can be observed, necessitating multiple tubes to provide any level of confidence that significant defects will be observed. Samman and O'Neill give no clear guidance on the number of tubes that would be appropriate, but recommend a larger number of tubes than would be necessary for Cross-Hole Sonic Logging (CSL) or Gamma–Gamma Logging (GGL). If more extensive examination of an anomalous zone is required, either CSL or GGL could be performed, provided that care is taken when installing the access tubes to ensure that they are large enough and straight enough to accommodate the appropriate transducers.

At the time the article was written, the depth of shaft that could be tested with the optical technique was limited to about 30 m (100 ft) by the available equipment. There appears to have been little follow-up on this work, and the present authors have been unable to find any references to commercial applications of the technique.

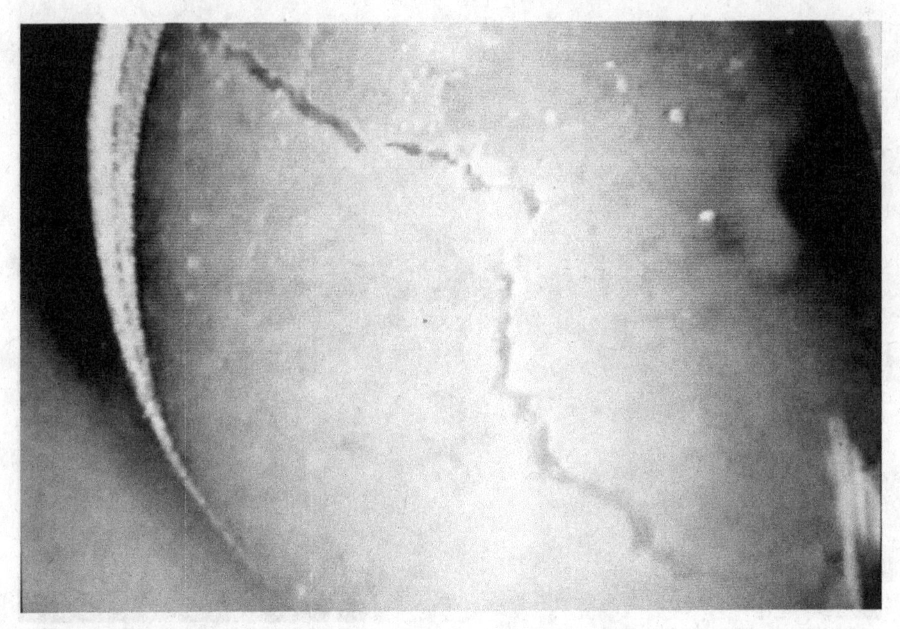

Figure 13.9 Forward view of a crack in the concrete. Reproduced by permission of Stress Engineering, Inc., Texas, USA

Figure 13.10 Lateral view of 'honeycomb' or leached concrete. Reproduced by permission of Stress Engineering, Inc., Texas, USA

13.4 GUIDED WAVE ANALYSIS

The simplest forms of wave propagation analysis in deep foundation shafts, such as those used in the conventional analysis of surface reflection methods, like Sonic Echo and Sonic Mobility, assume that the wave propagation is one-dimensional – along the axis of the shaft. For this condition to be true, the length of the stress wave must be greater than the diameter of the shaft, so that most of the wave energy is contained within the shaft with negligible leakage into the surrounding soil, and the wave can be considered 'non-dispersive'. This means that the wavelength must become greater as shaft diameter increases, which, since frequency is inversely related to wavelength, means in turn that the usable frequency range decreases as shaft diameter increases. The drawback to this approach is that waves with large wavelengths will only be reflected by significant changes in condition, such as large inclusions, discontinuities or the base of the shaft. Research and site experience in shafts with known defects has shown that smaller discontinuities and inclusions cannot be reliably identified (Baker *et al.*, 1993; Samman and O'Neill, 1997a).

A research team at Northwestern University, Illinois, under the guidance of Professor Richard Finno, has examined the relationship between wavelength and anomaly resolution, and the effect of the foundation shaft as a waveguide The Northwestern team derived the wave-frequency equation for concrete piles embedded in soil (Hannifah, 1999). Their work describes the wave energy as traveling along the shaft in a group or 'packet' of waves at many different frequencies. The velocity of the wave packet is regarded in the conventional methods of analysis of surface reflection techniques as the propagation velocity that is used to calculate the depth of the reflecting feature.

Finno's team has shown that stress waves become dispersive at wavelengths smaller than the shaft diameter – i.e. the waves contain significant components that travel in directions other than the axial propagation of the lower-frequency waves, and wave velocity becomes a function of wave frequency (Finno *et al.*, 2001). The radial component of the wave packet causes energy to leak out of the pile into the soil, thus attenuating the energy in the foundation shaft. The radial component becomes more important as frequency increases, thus also increasing the attenuation of wave energy in the shaft by an amount that is dependent on the stiffness of the soil.

A second critical factor is the assumption made in conventional analysis that wave propagation velocity is constant. According to Finno's team, propagation velocity is reasonably constant in the low-frequency range, where the wave is non-dispersive. At higher frequencies, however, as the wave becomes dispersive, propagation velocity decreases rapidly. Thus, the conventional analysis methods can only yield accurate results at relatively low frequencies. As shaft radius increases, the wavelength at which the wave becomes dispersive also increases, meaning that the frequency range in which the propagation velocity remains constant is reduced.

This research suggests that the attenuation of the wave energy and the change in wave propagation velocity limit the usable frequency range for surface reflection techniques to about 1000 Hz for shafts less than 1.0 m in diameter, dropping to about 600 Hz for shafts of 2.0 m diameter. However, in the present authors' experience, the useful frequency range in the Sonic-Mobility test extends well beyond 1000 Hz, even for large-diameter shafts. The upper frequency limit appears to be controlled rather by the useful energy spectrum generated by the impact hammer tip, and can be between 1.5 and 2 kHz, depending on the material that the hammer tip is composed of.

13.5 STATISTICAL ANALYSIS

The constant advances in nondestructive test methods continue to improve accuracy and economy, but do nothing to address a key question for engineers who are specifying test programs – how much testing is enough? How many tests need to be done in order to have confidence that no serious problems are being missed? The flip side of this for the project owner is how much money have we wasted on unnecessary tests? Research is continuing into the statistical reliability of applying NDT methods to quality control of new deep foundation construction (Baker *et al.*, 1993; Turner, 1997; Davis, 1999; Cameron *et al.*, 2003; Cameron and Chapman, 2004).

Williams and Stain (1987) addressed this issue with an empirical approach in their 1987 paper, based on site experience and the results of several thousand tests. They gave 'decision trees' that help to determine the appropriate number of tests for a given population of shafts (Figure 13.11) and to select an appropriate method

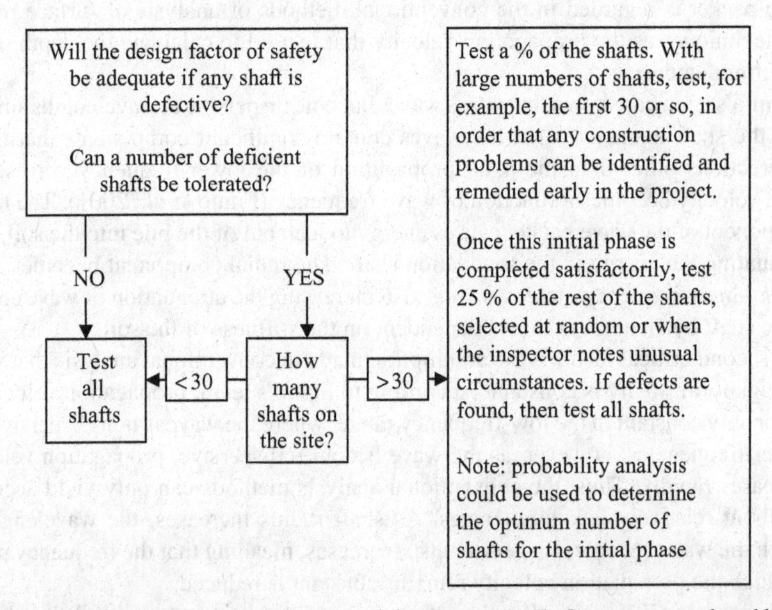

Figure 13.11 'Decision tree' for determining how many shafts to test reproduced by permission of Testconsult Ltd, Risley, UK

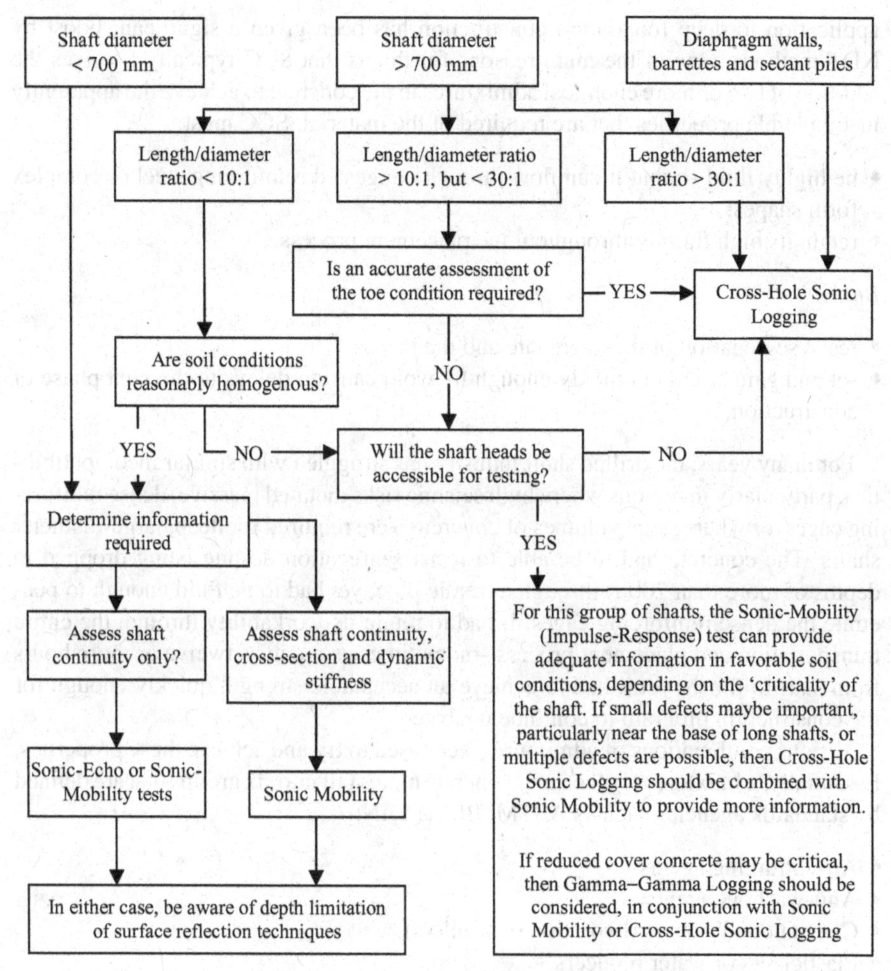

Figure 13.12 'Decision tree' for selection of test method reproduced with permission of Testconsult Ltd, Risley, UK

or combination of methods (Figure 13.12). More recently, several researchers have used probability theory to investigate this problem. The current state-of-the-art of this subject is discussed in Chapter 12 of this book and it is certain that it will continue to provide a fruitful avenue for study in the future.

13.6 SELF-CONSOLIDATING CONCRETE

The topic of self-consolidating concrete (SCC) may at first appear out of place in a book on advances in nondestructive testing, but in fact SCC and research into its application to deep foundation construction has been given a significant boost by NDT methods. One of the main reasons for this is that SCC typically involves the addition of two or more chemical admixtures to the concrete to achieve the apparently incompatible properties that are required of the material. SCC must:

- be highly fluid so that it can flow through congested reinforcing steel or complex form shapes;
- retain its high fluidity throughout the placement process.

But:

- resist segregation of the aggregate and the paste;
- set and gain strength quickly enough to avoid causing delays to the next phase of construction.

For many years, the drilled shaft industry has struggled with similar incompatibilities, particularly in regions where high seismic risks dictated massive, dense reinforcing cages, or where large volumes of concrete were required for deep, large-diameter shafts. The concrete had to be able to resist segregation despite being dropped to depths of more than 200 ft through a tremie pipe, yet had to be fluid enough to penetrate the dense reinforcing cages. It had to retain its workability through the entire transportation and placement process – sometimes as much as twelve or more hours from start to finish, yet it had to achieve an acceptable strength quickly enough for the construction program to continue unabated.

Various combinations of admixtures were used to try and achieve these properties. Essentially, admixtures for 'normal' concrete mixes fall into six groups that are defined by standards agencies such as ASTM, BSI or DIN:

- Air-entraining agents
- Anti-washout agents
- Cement conditioners (hydration controllers)
- Plasticizers or water reducers
- Set retarders
- Set accelerators.

Air entrainment is generally recognized as means of reducing the severity of freeze/thaw damage in exposed concrete, and may at first seem unnecessary in concrete for deep foundation work. A side effect of air entrainment, however, is to improve the viscosity of the cement paste, and thus reduce the risk of segregation – a valuable property in a high-slump concrete mix. In SCC mixes, resistance to segregation is improved by yet another class of admixture, called 'viscosity modifiers'.

Experience with multiple admixtures in deep foundation concrete has revealed several problems caused by unpredictable behavior of the concrete. In some cases, a high-slump mix suffered a sudden loss of slump when pumped through a tremie pipe under a high hydraulic head, such as a 200-ft-deep shaft filled with water or slurry.

Flash setting has occurred on numerous occasions – in some cases, so severely that the contractor has been unable to remove the tremie pipe, or pull a temporary casing.

A problem commonly identified in Cross-Hole Sonic Log (CSL) data in recent years is that the concrete in a shaft often does not set or mature uniformly. Some zones may set and gain strength quickly, while other zones mature much more slowly. The variability in the strength shows as variability in pulse velocity and amplitude in CSL data. The result is that numerous shafts have been rejected as containing anomalies, when in fact there was nothing wrong other than that the concrete in some portions of the shaft had already set and begun gaining strength, while the concrete in other parts of the shaft had not yet started to set. This variability is not yet fully explained, but experience has shown that it is more pronounced in large-diameter shafts that have both water-reducers and set retarders and/or cement conditioners. A number of cases have been documented where the concrete has taken more than two weeks to reach a uniform condition. In extreme cases, the concrete has taken more than a month to set! Ultimately, however, in each case the concrete reached a strength that met or exceeded the minimum specified.

These problems have been documented in various parts of the world. The present authors are aware of cases in several parts of the United States, including Alaska, plus Canada, Hong Kong and India. It is possible that small changes in mix proportions or water/cement ratio are the cause of the variability in some cases, but soil temperature and moisture content also appear to be factors. It also appears likely that there are synergistic effects with certain admixture combinations. The deep foundation industry is seeking solutions to these problems, and SCC appears to have potential for solving the placement and workability issues experienced with conventional concrete, but the use of multiple admixtures is presently still a cause for concern.

At the time of writing this book, construction industry experience with SCC has been primarily in the areas of flatwork, such as floors and pavements, and in walls where a narrow cross-section inhibited the flow of conventional concrete mixes. Recently, however, there have been some successful projects where SCC has been pumped into column forms from the bottom upwards. There is, at the time of writing, little experience with placement of SCC in deep shafts or mass concrete pours. The potential for excessive bleeding in such conditions has therefore not been evaluated, but a good start has recently been made in this direction by a team at Auburn University, Auburn, Alabama, under the guidance of Professor Dan Brown.

In an article published in *Foundation Drilling* journal (Brown, 2003), he describes the construction of five drilled shafts – two with conventional concrete containing 19 mm (0.75 in) maximum-size aggregate, two with self-consolidating concrete and one with a conventional 'pea-gravel' concrete mix. The shafts were exhumed after the concrete had set. Core samples were taken from the sides and each shaft was sliced into four pieces for visual examination. Laboratory tests on the concrete samples had not been completed at the time of writing the article, but the visual impression was that the SCC and pea-gravel mixes created a denser concrete outside the reinforcing cage than the conventional 19 mm aggregate mix. Professor Brown also noted increased air entrainment in the SCC mixes, resulting in increased porosity and some loss of strength.

One of the conclusions from this work was that we are still learning about proper use and interaction of admixtures. The present authors are active on committees of the ASTM and ACI and are currently championing research into these applications.

13.7 ACCEPTABLE VIBRATION LEVELS

There has been much concern about the effect of ground-borne vibrations on freshly placed concrete during the setting process. Engineers have been concerned that excessive vibration from other construction activities such as casing vibration or pile-driving may cause unacceptable cracking in nearby shafts during the setting process, when the strength of the concrete is low. Despite the engineers' concern, there has been very little research into the effect of ground-borne vibrations on hardening concrete embedded in the soil and so, to be safe, engineers generally set very conservative vibration limits. As a result, construction specifications often contain very restrictive requirements for work in the vicinity of newly constructed shafts. Some require vibrations to be limited to less than a certain level – typically 13 mm s^{-1} (0.5 in s^{-1}) – until the concrete has reached the specified strength, while others simply state that no vibration is permitted within a specified radius of the shaft for a certain number of days. Such restrictions can cause substantial delays, and hence increased cost, in the construction program.

A recent research program performed in Hong Kong, however, should help alleviate this problem. A paper published in the November–December issue of the *ACI Material Journal* (Kwan *et al.*, 2005) describes the results of research into the effects of shock vibrations on the integrity of 'early-age' concrete. Essentially, this key findings of this published research are that maturing concrete is not nearly as susceptible to vibration damage as was generally thought, and that the compressive strength of concrete was the least sensitive parameter, and therefore shafts that would be loaded in simple axial compression carried the least risk of significant damage.

This paper gives a set of recommended vibration limits that are significantly less restrictive than most currently used limits, and so offers the potential for more efficient use of construction time and equipment. Given the number of times that the present authors have been called to perform integrity tests on shafts that engineers had vibration-related concerns about, this publication should generate widespread interest, and, hopefully, some changes in specifications relating to allowable vibration levels.

13.8 AUTOMATED MONITORING SYSTEMS

Another topic that may seem out of place in this book is that of 'automated construction monitoring systems'. However, recent developments in the augered, cast-in-place

(ACIP) pile industry have shown that the automated monitoring systems currently on the market provide valuable warning of potential problems during the construction process, but can still leave the engineer with the problem of assessing the extent and severity of the recorded anomaly, and its likely effect on the working performance of the foundation. Thus, automated construction monitoring systems and nondestructive testing can form two complementary parts of the quality control process.

At the time of writing this manual, there are several proprietary construction monitoring systems for ACIP piles in use in various parts of the world, but all share several significant features. Sensors mounted on the drilling rig monitor several critical parameters, such as auger rotation speed, auger withdrawal rate, grout or concrete flow rate and pressure, and total volume of grout or concrete placed. All of these data are recorded continuously by a computer, which generates a graphic output in the control cab of the drilling rig to show the operator the shape of the shaft that is being constructed. In the event that the monitoring system shows evidence of a severe neck-in, the operator may have the option to reinsert the auger and re-drill the hole, effectively starting the construction process over again.

If soil conditions are such that re-drilling may cause more problems rather than fixing the neck-in, some form of additional testing may be called for to help assess the significance of the problem. Similarly, where reinforcing steel is inserted into the shaft after grout or concrete placement and unexpected resistance or obstruction is encountered, some form of NDT may be required to determine whether soil 'squeeze' or cave-in may have caused the problem.

This approach has been so successful in improving the quality of ACIP piles that it has made a significant contribution to the acceptance of the ACIP construction technique by all but the most skeptical of engineers or owners. This fact has not been lost on the drilled shaft construction industry, and two of the leading associations for drilled shaft constructors, the International Association for Foundation Drilling (ADSC) and the Deep Foundations Institute, are currently seeking research proposals for the creation of an automated monitoring system for drilled shaft construction.

13.9 WIRELESS ACQUISITION SYSTEMS

One of the major drawbacks of all nondestructive testing systems to date has been the need for access to the shaft head with either a fairly large data acquisition and recording system, or an 'umbilical' cable connection from the shaft head instrumentation to the recording system somewhere nearby. This problem is particularly exacerbated by difficult access conditions, such as in deep excavations, or over water to individual bridge or pier foundations.

Several manufacturers of foundation testing equipment have responded to these problems by producing 'portable' self-contained data-acquisition systems. The word 'portable' is used with some reservation, because the present state of battery technology causes a significant trade-off between working times and the size and weight of the system. Some manufacturers have opted for a relatively short battery life, with replacement units on charge elsewhere on the site, while others have opted for larger, heavier batteries which provide almost a full day of capacity, but add a substantial amount to the weight of the system, making it awkward to carry around congested, uneven sites or up ladders, etc.

Pile Dynamics, Incorporated (PDI), of Cleveland, Ohio, provided one clue as to the future of foundation testing equipment in 2002. PDI supplies high- and low-strain testing equipment for deep foundations, and has recently introduced a pile-driving monitoring system where the test data are relayed directly from the site to the analyzing engineer by cellular telephone (Frazier *et al.*, 2002). Thus the engineer can concentrate on the analysis in the relative comfort of an office, free from the typical noise and distractions of a busy construction site.

With the current rapid evolution of handheld computers and further miniaturization of data-acquisition systems, it is realistic to expect *truly portable* data-acquisition systems capable of wireless communication with computers for data storage and analysis, either on-site or in remote office locations within the next few years.

13.10 'SMART' STRUCTURES

Research funded recently by the Florida Department of Transportation (FDoT) took a slightly different approach, resulting in the creation of 'smart' piles. These pre-cast piles were instrumented with strain gauges and accelerometers cast into the concrete at both the top and the base of each pile, allowing the forces generated within the pile during driving to be measured much more accurately than with the traditional high-strain dynamic test methods (McVay *et al.*, 2004).

The instrumentation is self-contained, with a battery and radio transmitter for wireless transmission of the data to a receiver connected to a laptop computer. The instrumentation package can be programed to conserve battery power by going into a 'sleep' mode after a pre-set time interval, until it is 'woken up' again by a signal from the computer. This facility allows additional measurements to be made at selected intervals after driving to evaluate changes in conditions, such as dissipation of residual stresses, or the distribution of load as the superstructure is constructed.

One of the most significant advances made in this research has been the development of a relatively inexpensive and durable set of transducers. Cost is an important factor, since the instrumentation package is entombed in the foundation forever, or, at the very least, for the life of the structure, and so is regarded as a sacrificial, or 'lost' package. 'Technology transfer' is the politically correct term used today for the practice of 'borrowing' ideas from other walks of life or branches of engineering. In this case, accelerometers and strain gauges developed for mass-production in the automotive and aviation industries, and microelectronics developed for the cellular telephone industry, provided the inexpensive and rugged components needed for the 'lost' instrumentation packages.

This technology can be utilized readily by the drilled shaft industry to evaluate shafts during high-strain or rapid-load testing, and in the early stages of service life to evaluate performance under load. Hopefully, the information gained will enable us to validate common assumptions, debunk a few myths and calculate the true factor of safety, allowing a less conservative approach to foundation design. In short – the knowledge gained from a few such projects may have a profound effect on the design and construction practices for deep foundations.

14

The Place of Nondestructive Testing at the Beginning of the 21st Century

After nearly forty years of nondestructive testing applied to deep foundations in one way or another, it is pertinent to discuss the role played by this technology in the construction industry throughout the world.

The first question to be asked is: *Why, and by whom, is nondestructive testing needed*? It is interesting that the first sponsors of NDT research and development in the 1960s and early 1970s were piling contractors, through research organizations such as CEBTP in France and TNO in Holland. As the early methods began to show some promise, consulting engineers and public owners quickly saw the potential advantages in applying these techniques to site conditions (Levy, 1970; Gardner and Moses, 1973; Preiss, 1971; Baker and Kahn, 1971). It must be said that testing in the first decade (1967–1977) was still regarded in many engineering quarters as unproven, with too much subjectivity in test-result interpretation. Testing organizations were learning progressively about the capabilities and limitations of the various methods as site testing was introduced, and it was difficult to accept that NDT could play a contractual role in the acceptance or rejection of any part of a piled foundation. No official specifications, recommendations or standards were drawn up until 1978, when the French private construction industry (DTU, 1978) and the Greater London Council, UK each wrote recommendations for the inclusion of NDT in piling contracts.

This development saw the arrival of purely commercial testing organizations on the scene in France, Holland and the UK, as well as in parts of the world where these three countries influenced the construction industry (Hong Kong, Singapore, Egypt, former French Africa and the former French Caribbean Islands, for example).

Nondestructive Testing of Deep Foundations B.H. Hertlein and A.G. Davis

Testing specifiers still came from different sectors of the industry, such as public owners in housing and transportation networks, architects and engineers representing private owners, and in one exceptional case, one large piling contractor in the UK that purchased testing equipment to 'auto-control' its own construction projects. At this time, in the present authors' opinion, one unfortunate development was the trend to include the NDT contract in the main piling contract. This meant that the contractor managed the testing contract (usually the site project manager) with items such as selection and timing of piles to be tested and payment to the test contractor under his or her direct control. Reporting of test results under this system was directly to the contractor, often with express instructions to not report to or to discuss the test results directly with the owner or the owner's representative. 'The fox was being paid to guard the hen-house!'.

The USA still remained a part of the world that had to be convinced about the place of NDT in the quality control of deep foundations. Acceptance of the available test methods in the 1980s was very slow and sporadic, and was limited to special cases where 'forensic-style' examination of problem foundations was required. However, with the advent of larger, single foundations for bridge piers provided by large-diameter drilled shafts, particularly in areas with high seismic risk, those involved in public-sector transportation engineering, led by the FHwA, saw that some form of quality control incorporating NDT would be advantageous. A test program sponsored by the FHwA in 1989 through 1991 examined the suitability of available low- and high-strain test methods for the problems faced in the control of drilled shafts (Baker *et al.*, 1993). The 'decision tree' developed by Baker *et al.* stressed the advantage of down-hole over surface tests in satisfying quality control needs for single, large-diameter shafts.

This recommendation instilled doubt in USA engineers about the usefulness of surface testing, and their inclusion in specifications drawn up by public-sector engineers became less common. It is interesting to note that the European approach to method selection, as expressed in the 'decision trees' shown in Chapter 13 (Figures 13.11 and 13.12) (after Williams and Stain, 1987), is more open to the acceptance of surface testing when used to control the quality of relatively large pile groups. The American viewpoint will probably change now that ASTM standards have been written for some of the low-strain integrity tests. However, the recently developed ASTM 'Standard Test Method for Low-Strain Integrity Testing of Piles' (ASTM, 2000a) incorporates quite different test methods, without any commentary as to their respective advantages and disadvantages.

The Canadian experience with quality management is particularly relevant at this juncture – not just for deep foundations, but for the construction industry as a whole. A paper entitled 'Achieving Excellence in Canadian Construction', presented by Wennerstrom to the American Society for Quality (Wennerstrom, 2004), is particularly eloquent on the subject. Wennerstrom discusses the quality management concept and presents two case histories that illustrate the effectiveness of an integrated approach to quality management, rather than the typical program of piece-meal efforts by various subcontractors that is so prevalent today.

Two of the salient points in Wennerstrom's paper are first, that rework, or remediation of errors, is one of the leading factors in driving up the cost of construction in Canada and the United States, and secondly, that the construction industries of both countries have fallen well behind their economic competitors in several other countries in the application of quality management programs such as those laid out in ISO 9001-2000, published by the International Organization for Standardization (ISO). Wennerstrom shows that since the early 1990s, 597 construction contractors achieved ISO 9001 registration in Hong Kong, and 246 in Singapore, but only 192 in Canada, and less than 500 in the United States. A similar pattern is evident in Europe, with more than 80 % of UK contractors registered. Some of the resistance toward registration is the rumored complexity of the process, and the perception that ISO 9001 is written for manufacturing industries, and is therefore not applicable to construction. This latter argument is fallacious, not only because ISO 9001 is a generic quality process, and is therefore not industry-specific, but also because construction is, essentially, a manufacturing process. A combination of raw and pre-processed materials is taken and assembled to form the desired product, be it a bridge, a tunnel, a dam or an office.

Based on the concepts in ISO 9001-2000, Wennerstrom points out that the quality management process should not only be monitoring the quality of work being performed, but also driving continual improvement in the construction processes. Nondestructive testing of both the foundations and the superstructure is of particular importance to such a program, because the early feedback provided by the test results enables the contractor to modify and improve construction methods 'on-the-fly' and so gain maximum advantage from NDT techniques.

It is worth making a brief detour here to explain why the International Organization for Standards is called 'ISO', instead of 'IOS'. It was created from the outset to be an international organization, to set standards that would be equally applicable in all countries. The name, therefore, had to translate into different languages, or otherwise the names of standards would vary – IOS-9000 in English would be OIN-9000 in French (for Organization Internationale de Normalisation) or ION-9000 in German (for Internationale Organisation für Normung). The name ISO was derived from the Greek word 'Isos', which means 'Equal'. Thus, the abbreviated form of the name is always ISO, no matter which language is being used. ISO is a voluntary organization, and each participating member country has one vote, regardless of size or economic clout. The standards developed by ISO therefore represent the democratic voluntary consensus of the participants. Although they are voluntary, some ISO standards are referenced in certain national legislation or government specifications, making them mandatory in such applications.

Hopefully reading this book has left the reader with an understanding of the capabilities and limitations of the various nondestructive test methods, and the importance of selecting a method or program of methods that are appropriate to both the site and access conditions, and to the information that is required. Having said that, there are a few other points that should be considered before making a final selection.

14.1 NONDESTRUCTIVE TESTING AND LOAD AND RESISTENCE FACTOR DESIGN

An important change has occurred in engineering design philosophy in the last decade, at least in the United States and several other western countries. Arguably, 'Load and Resistance Factor Design' (LRFD) has been around since the middle of the 20th Century and has been used extensively in Europe for structural design, but it has only recently come into favor with influential engineering groups in the Americas, such as the American Association of State Highway and Transportation Officials (AASHTO), the American Concrete Institute (ACI) and the American Institute of Steel Construction (AISC). It has, however, gained favor rapidly. By 2000, LRFD was so widely accepted in North America that codes had been written or were being prepared for concrete construction, engineered wood construction, masonry construction, steel construction and timber construction.

The deep foundations industry was no exception. LRFD bridge foundation design was first implemented in a few provinces of Canada in the 1980s. The United States were somewhat slower in embracing LRFD for foundations, but it is now specified by the FHwA, and will be mandatory on all FHwA-sponsored projects by 2007. Most state departments of transportation in the United States will no doubt follow the FHwA's lead for bridge and bridge foundation design. Among engineers who have been taught 'Allowable Stress Design' (ASD) methods, LRFD is often seen as a means of saving costs by 'shaving' safety margins – an approach that seems reasonable, provided that all of the factors involved are closely controlled. That is where NDT methods become important quality assurance tools and design aids, as noted in the Foreword to this book. A fact that is sometimes overlooked by designers and specifiers is that the LRFD approach assumes that there is a comprehensive and well-managed quality assurance program throughout the construction process, which permits the acceptable margin of safety of the design to be *rationally* reduced to save costs. The experience of the Hong Kong Housing Authority in Shatin in the late 1990s, as described in Chapter 1 – Introduction and a Brief History, is a clear demonstration that having correctly completed paperwork is meaningless if the inspection and testing programs have not been conscientiously and accurately performed (Hong Kong Housing Authority, 2000).

Further information on the recommended practice of LRFD for deep foundations in the United States is available in the form of a Transportation Research Board (TRB) report from the National Cooperative Highway Research Program (NCHRP, 2004). NCHRP Report No. 507, for Research Project No. 24-17, published in August, 2004, is entitled 'Load and Resistance Factor Design (LRFD) for Deep Foundations'.

This report examines resistance factors for both drilled shafts and driven piles and recommends procedures for calibrating deep foundation resistance for use in an LRFD program.

14.2 SETTING UP AN EFFECTIVE QUALITY MANAGEMENT PROGRAM

An effective quality management (Q/M) program for deep foundations must begin at the design stage and follow through all aspects of the construction program. It cannot be stressed enough that nondestructive tests for deep foundations were never intended to be a substitute for a competent inspector. The present authors have seen a distressing number of projects in which reliance for quality control was placed on the testing program instead of on a competent inspector. Specifying a program of nondestructive tests does not imply that money can be saved by cutting back on inspection – the real intent should be that the data acquired by the NDT program will support the inspector by providing additional information for those instances when the inspector notices something amiss. A good Q/M program will include the following:

- Selection of a qualified and experienced deep foundation contractor.
- Selection of an experienced construction inspector, preferably from the geotechnical firm that designed the foundations, or at least performed the soil exploration borings on the site.
- Selection of an appropriate program of nondestructive quality assurance tests.
- Selection of an experienced testing company to perform the nondestructive tests.

Although the foundation inspector's role was discussed in Chapter 4, the salient points bear repeating at this juncture. The inspector is of particular importance to the Q/M program, because if an anomaly is located by nondestructive testing, the inspector's notes will most probably provide critically important clues to help the foundation contractor and the engineer determine the cause and the nature of the anomaly, and hopefully figure out appropriate measures to prevent it from reoccurring. In order to take adequate and accurate notes, the inspector must be experienced enough to understand the critical factors and recognize deviations from the designed or anticipated conditions and procedures. The critical factors that the inspector must be familiar with are:

- Site soil conditions.
- The design philosophy for the foundation (whether the foundation is designed to support its load in end-bearing, side friction, or a combination of the two).
- The proposed construction procedure and equipment.
- The concrete mix design requirements (slump/flow and/or duration of workability) and the factors that affect its workability and stability, such as premature slump loss, bleeding and segregation.

The inspector must guard against developing an adversarial relationship with the contractor. Without the contractor's full and willing cooperation, it is virtually impossible to run a complete and thorough inspection program. Assuming that the contractor, inspector and tester are all experienced, competent and cooperative individuals, then, once all of the information from the soil exploration borings, the contractor's records, the inspector's notes and the nondestructive test results have been compiled and analyzed, the engineer is usually in a strong position to determine the most likely cause of the anomaly. It becomes a process of elimination, in which impossibilities and unlikely events are gradually whittled away until the most logical answer is defined. The conclusion sometimes results in declarations of incredulity on the part of the contractor, and sometimes on the part of the inspector too, but, to paraphrase Sir Arthur Conan Doyle's most famous character, Sherlock Holmes, 'Once you have eliminated the impossible, then whatever remains, however improbable, must be the truth'.

14.3 WHO'S TESTING THE TESTER?

Just like the foundation inspector, the nondestructive testing personnel must guard against letting adversarial relationships develop with the contractor and field crew. There are many ways in which the contractor's crew and workload can affect the ease of access to the foundations, the performance of the NDT and the quality of the resulting data. An uncooperative contractor can make the testing difficult at best, and inconclusive or misleading at worst, drastically reducing its value to the Q/M program.

No amount of nondestructive testing will provide quality assurance if the test operator is inadequately trained or is not experienced enough to recognize bad data when it occurs. Unfortunately for the owners and specifiers of deep foundation projects, there are, at the time of writing this book, no internationally recognized standards or curricula for training the technical personnel that perform low-strain nondestructive testing. There is a wide diversity of education and experience among testing personnel. Some firms hire only registered engineers to perform the work – others train field technicians. In the present authors' experience, the educational background and professional status of the tester is of little importance – it is the quality of the specific training that he or she has received that matters most. An engineer with a Ph.D. will make mistakes if he or she has not been adequately trained in the specifics of the test method being employed. Most firms that sell nondestructive testing equipment offer some training in its use, but the duration and quality of that training is highly variable. The biggest factor determining the competence of most testing personnel is the quality of the training and mentoring that they have received 'on the job' from their employers or supervisors.

Of course, before there can be agreement on the competence of a particular person to perform a specific test, there must be agreement on the way in which that test should be performed. The national standards organizations are a vital part of this process. In the United States, there has been an ASTM standard practice for high-strain testing since 1986 – ASTM D4945, 'Standard Test Method for High-Strain Dynamic Testing of Piles', (ASTM, 2000b), but the standards for Impulse-Echo or Impulse-Response testing and for Cross-Hole Sonic Logging have only recently been approved – ASTM D5882-00, 'Standard Test Method for Low-Strain Integrity Testing of Piles' in 2000 (ASTM, 2000a) and ASTM D6760, 'Standard Test Method for Integrity Testing of Deep Foundations by Cross-Hole Testing' in 2002 (ASTM, 2000). There is, as yet, no internationally accepted standard for either Parallel-Seismic testing or Gamma–Gamma Logging, although the French organization AFNOR does have a standard for both methods. AFNOR has been a leader in the area of standardization for NDT methods, having published standards for the Impulse-Echo and Parallel-Seismic tests in 1993, and for the Impedance (Impulse-Response) test in 1994. The California Department of Transportation (CALTRANS) is the only regular user of the Gamma–Gamma method in the USA, and has its own guidelines for using the method and interpreting the data, but these have not been officially adopted for wider application. The present authors have been involved in projects in Nevada and Arizona, however, where the Gamma–Gamma method was used, in both instances following the CALTRANS practice.

Testing standards often fall victim to national pride. Some countries readily accept the benefits of technology transfer – others insist on doing things their own way. Until a country decides to either adopt a standard nondestructive testing practice from another country, or to develop its own, the examination and certification of nondestructive testing personnel in that country will be 'left in limbo'. In the UK for example, there are no British Standards for nondestructive testing of foundations. The only official publication that even resembles a guideline is a brief discussion of testing in the Institute of Civil Engineers (ICE) 'Specification for Piling and Embedded Retaining Walls' which was published in 1996. Australians are in similar straits – Australian Standard AS2159-1995, 'Piling – Design and Construction', provides some useful guidelines for high-strain testing but virtually ignores the low-strain tests. In Japan, the Japanese Industrial Standards Committee (JISC) publishes Japanese Industrial Standards through the Japanese Standards Association (JSA). The JISC Website lists available standards and acknowledges that international standards may be adopted whole or in part, but no standards for deep foundation testing were listed at the time of writing this book.

An excellent review of the state of the world's standards for foundation testing was published at the 1998 DFI conference by several employees and associates of Pile Dynamics, Inc. (Beim *et al.*, 1998). This review listed only eight countries that had established either standard procedures or codes of practice that recognized nondestructive tests for deep foundation, and even those were largely geared to high-strain testing. A few other countries had some form of recommendation or unofficial document published by engineering professionals, but no official standards.

Beim *et al.* (1998) also reported that many countries allowed or required the use of dynamic test methods, even though they had no standards or codes of practice for the engineering profession to refer to. In Sweden, for example, Beim *et al.* (1998) reported that there are no national standards for foundation testing, but the Swedish Commission on Pile Research had published some guidelines for high-strain dynamic and static-load testing, The Swedish guidelines made no mention of low-strain tests. Enquiries made by the authors of this present book with the secretary of the Swedish Commission on Piling in early 2004 confirmed that the situation there had not changed significantly.

The review by Beim *et al.* (1998) stated that, at the time it was written, some other countries had standards, but their building or construction codes did not reference the standards, nor did they mandate their use. The countries that are currently identified as having produced practical standards that were usually embraced by building or construction codes were identified as shown in Table 14.1, but there are some 'oddities' there too. Canada, for example, in 2000 adopted the former 'Ontario Highway Bridge Design Code', which includes the requirements for high-strain dynamic testing, as 'National Bridge Design Code CAN/CSA S6-00', yet the Canadian Building Code is a performance-based document, which requires only that the foundation be safe and provide a serviceable support for the structure. The choice of quality assurance and test methods and the way in which they are performed is left entirely up to the engineer.

In the Far East, it seems that several countries are presently using, or propose to adopt, the ASTM Standards. From personal correspondence with engineers from Thailand and Malaysia, the present authors understand that, as of Spring, 2004, engineers in Thailand and Vietnam use ASTM standards, although the Engineering Institute of Thailand is working on the Thai Industrial Standard (TIS) which will include nondestructive testing of foundations (Hertlein, 2004c). In China, the China National Institute for Standardization (CNIS) was formed in 1999. CNIS is in fact an umbrella organization, comprised of The Institute of Standardization Theory and Strategy, The Institute of Basic Standards, The Institute of Resources and Environmental Standards, The Institute of Information Technology Standards and the Institute of Quality Control Standards. Each of these institutes is in the process of adopting ASTM standards where applicable, or creating new standards where necessary (Weihua, 2004).

At the time of writing this manual, the field of high-strain testing is better served than low-strain testing in the area of standardization and education. At least two manufacturers of high-strain testing equipment offer training courses and ongoing 'on-the-job' consultation that are officially recognized by their respective national education accreditors and are geared to a series of professional competence examinations that have been devised by Foundation QA of Croydon, Victoria, Australia, and are supported and/or administered worldwide by Pile Driving Contractor's Association.

The PDCA/Foundation QA examination is a multiple-choice exam in two parts. Part A is concerned with the data-acquisition procedure and examines the operator's level

Table 14.1 Standards and/or codes for practice for nondestructive foundation tests.

Country	Method	Reference	Title
Australia	High-strain	AS 2159-1995	Piling – Design and Installation
Brazil	High-strain	NBR-13208	Dynamic Testing of Piles
Canada	High-strain	CAN/CSA S6-00	Canadian Bridge Design Code (formerly Ontario Highway Bridge Design Code, 3rd Edition)
China	High-strain	JGJ 106-97	Specification for High-Strain Dynamic Testing of Piles
	Impulse Echo and CSL	JGJ 94-94	Technical Code for Building Pile Foundations: Chapter 9.1, Quality Inspection of Pile Installation
France	CSL	NFP94-160-1	Sols: Reconnaissance et Essais: Auscultation d'un Élément de Fondation – Partie 1: Méthode par Transparence
	Impulse Echo	NFP94-160-2	Sols, etc. – Partie 2: Méthode par Réflexion
	Parallel Seismic	NFP94-160-4	Sols, etc. – Partie 3: Méthode Sismique Parallèle
	Impulse Response	NFP94-160-4	Sols, etc. – Partie 4: Méthode par Impédance
	Gamma–Gamma Logging	XP94-160-5	Sols, etc. – Partie 5: Méthodes par Diffusion Nucleaire à Rayonnement Gamma
Germany	High-strain	DGGT Working Group 2.1 (1998)	Empfehlungen des Arbeitskreises 2.1 der DGGT für statische und dynamische Pfahlprüfungen
UK	High-strain, CSL, Impulse Echo and Impulse Response	Institution of Civil Engineers	Specification for Piling and Embedded Retaining Walls: Specification, Contract Documentation and Measurement (Guidance Notes)
USA	High-strain	ASTM D4945	Standard Test Method for Dynamic Testing of Piles
	Impulse Echo and Impulse Response	ASTM D5882	Standard Test Method for Low-strain Integrity Testing of Piles
	CSL	ASTM D6760	Standard Test Method for Integrity Testing of Deep Foundations by Cross-Hole Testing

of knowledge concerning material properties correct procedures and potential data quality problems. Part B examines the ability of the operator to make preliminary analyses of the test data in the field to identify possible driving problems, detect damage, offer guidance to the pile-driving contractor or the engineer and determine pile capacity.

The examination results are graded into six categories, effective as of January 2004:

- < 50 % – Fail, not certified.
- 50–64 % – Lower Basic certificate.
- 65–75 % – Upper Basic certificate.
- 75–84 % – Lower Advanced certificate.
- 85–92 % – Upper Advanced certificate.
- 93–100 % – Expert certificate.

All certificates are valid for a period of five years from the date of issue. The examinee may retake the examination at any time before the certificate expiry date in order to improve his or her certification level. The period of certificate validity, however, has been the cause of considerable controversy in the high-strain testing community, with many people pointing out that, for example, a licensed civil engineer does not have to re-take his certification test every five years – why should a testing engineer? This thorny topic is, at the time of writing, being reviewed by the Testing and Evaluation Committee and the Trustees of the DFI who formerly sponsored the exam.

Regardless of the outcome of the DFI Trustees' deliberations, it is to be hoped that international recognition of the PDCA Foundation QA certification or its successor will set a standard for high-strain testers that will eventually be emulated by those in the low-strain testing field. It will require the establishment of nationally or internationally accepted standards for the test methods, recognition by employers that expert training and a commitment to ongoing professional development is required, recognition by manufacturers of nondestructive test equipment that the training they offer must comply with the pertinent national or international standards and recognition by the testing personnel that they must make an investment in their personal careers by taking advantage of the training available to them, and above all – 'doing their homework'!

14.4 ACCEPTANCE CRITERIA

Another aspect of NDT that is often mentioned by foundation contractors is the lack of consistency among engineers when dealing with anomalous shafts. The determination of whether an anomalous shaft is acceptable or not depends very heavily on the experience of the engineer, and his or her familiarity with both the shaft construction method and the test method, and understanding of the design philosophy for the shaft. Many experienced engineers have long-believed that the significance of a particular anomaly can vary from substantial to negligible, depending where it is located within the length of the shaft, but at meetings of the Deep Foundations Institute and the ADSC-IAFD, foundation contractors have reported widely differing responses to essentially similar anomalies.

At the time of writing this manual, the Testing and Evaluation Committee of the Deep Foundations Institute is discussing the possibility of preparing some guidelines for drilled shaft evaluation criteria, based on the results of Cross-Hole Sonic Log testing, but this has turned out to be a thorny and controversial topic and so it is likely to be some time before there is enough consensus of opinion to put anything in writing.

14.5 EVALUATING DEFECTS

A research program that will, hopefully, make an important contribution to the drilled shaft evaluation and acceptance process is, at the time of writing, in the second of three proposed phases. The research program is being conducted by Katherine Petek at the University of Washington, under the guidance of Professor Robert Holtz and sponsor Conrad Felice of Lachel, Felice and Company. The first phase of the work – analysis of stress distribution as a result of defect location and soil properties – was published in Petek's MSCE Thesis in 2001 (Petek, 2001) and in 2002 at the ASCE Geotechnical Conference (Petek *et al.*, 2002).

For the first phase of this research, a two-dimensional model was developed using a finite element program that allowed various shaft/soil combinations to be analyzed via a numerical load test. In this work, Petek showed that under axial loading, the significance of a defect is dependent upon the strength of the soils around the shaft. A shaft in stiff soil is usually designed to take greater loads than a shaft in weak soil. Therefore, a defect in a shaft in stiff soil will have a greater effect on the shaft's capacity than the same defect would have on a shaft in weak soil.

The position of the defect is also of great importance to the capacity of the shaft. Under axial loading, a defect near the head of the shaft will have more effect than the same defect would have if it were lower down the shaft. Petek reported that defects in the lower half of the shaft had a relatively small effect on shaft capacity, compared with defect(s) near the top. In fact, in many cases, it was the soil strength that limited the shaft load, even in defective shafts.

The overall research program was described in an article in the *Foundation* journal (Felice *et al.*, 2003). The second phase of this research program will assess three-dimensional models of drilled shafts that are subjected to axial and lateral loads. As in the first phase of the work, 'perfect' shafts will be modeled first and calibrated against real load-test data. Shafts with defects in them will then be modeled and the models validated by comparison with load-test data from real shafts with deliberately created defects.

The third phase proposed for this research will then attempt to assess the effects of defects in shafts subjected to dynamic loads such as traffic, wind-loading or earthquake conditions.

Appendix I Stress-Wave Propagation in Cylindrical Structures

1. GENERAL THEORY

Stress-wave propagation methods use the principle of the propagation of seismic waves in cylindrical or prismatic structures. Wave propagation in these structures is dispersive, which means that the phase velocity depends on the frequency and that several propagation modes can exist. Fortunately, most pile elements to be tested have high length-to-diameter (l/d) ratios, often from 10 to 40. The frequencies used in the Sonic-Echo and Vibration tests to obtain response from depth are low and the corresponding wavelengths are great compared to pile diameters.

Take the example of a 20-m long pile ($l = 20$) with a concrete wave propagation velocity, v_c of 4000 m s^{-1}. Then, $\lambda = 4l = 80$ and the resonant frequency, $f = 4000/\lambda = 50$ Hz.

If several successive frequencies are required to be seen, the sampling frequency can be increased to 500 Hz, say, equivalent to an 8-m wavelength, still large with respect to the pile diameter. This formed the basis for the introduction of the Vibration method.

For the Sonic-Echo method as practiced in the 1970s with analog signal techniques, conditions were less favorable. Take, for example, a break in a pile at a depth of 12 m; the time for the bar-wave signal to return to the pile head is $24/4000 = 6$ ms. If the depth is to be defined with an accuracy of $\pm$ 5 %, the echo position must be located at $\pm$ 0.3 m, which requires a band pass frequency reaching approximately 1000 Hz.

Soil damping has to be considered in all pile head stress-wave methods, and wavelengths in the soil must be assumed to be much greater than the pile diameter. This holds true for both compression and shear waves. Take, for example, a soil with a

Nondestructive Testing of Deep Foundations B.H. Hertlein and A.G. Davis

compression wave velocity of 1500 m s^{-1} and a shear wave velocity of 800 m s^{-1}. At 500 Hz, the wavelength is 1.6 m, close to the diameter of the tested pile. As a result, calculated damping will be only an approximation at high frequencies.

INFLUENCE OF DAMPING BY THE LATERAL SOIL

Under the influence of pile lateral surface displacement, the surrounding soil will move, creating waves with cylindrical symmetry. Their apparent wave propagation velocity along the z-axis will be equal to the pile wave velocity. As the soil wave velocity at the pile/soil interface is generally considerably less than that in the pile, the resultant wave directions in the soil are at a large angle to the vertical, into the soil. Total refraction can occur with emerging angles equal to the total refraction angles. A considerable part of the energy is radiated into the soil, and the energy transmitted by the pile is damped.

A good approximation of this phenomenon can be made, especially for the case where the soil is considerably less stiff than the pile, by following the classical method for the losses in channeled waves. Here, it is assumed that:

- The presence of soil does not perturb pile surface movements.
- Waves produced in the soil by the pile surface movement can be calculated.
- Energy radiated into the soil per unit pile length can be calculated. This radiated energy provides a measurement of the damping per unit pile length.[1]

The vertical elastic reaction from the surrounding soil is also calculated.

GENERAL CALCULATION METHOD

Cylindrical coordinates r, z will be used; z is taken as the pile axis coordinate and r is the radial coordinate. Displacement along z is w and displacement along r is q.

$\omega = 2\pi f$ (angular frequency)
$\lambda, \mu =$ Lamé's constants
$c =$ velocity of wave propagation along z-axis
$E =$ modulus of elasticity
$\rho =$ density
$\alpha = \sqrt{(\lambda + 2\mu/\rho)}$ (velocity of longitudinal waves)
$\beta = \sqrt{(\mu/\rho)}$ (velocity of transverse (shear) waves)
$v = \omega/c$
$k_\alpha = \omega/\alpha$

[1] This method does not allow the calculation of the modification of the phase velocity by the soil. This velocity is assumed to be between the bar-wave velocity (free pile sides) and the longitudinal velocity obtained if the pile were to be buried in the same material (semi-infinite medium).

$k_\beta = \omega/\beta$
$k^2 = k_\alpha^2 - \upsilon^2$
$k_1^2 = k_\beta^2 - \upsilon^2$
p_{rr}, p_{rz}, p_{zz} = stresses
S = section area
Z = resistance (impedance)
Y = admittance (mobility)
Z_c = characteristic impedance
Z_∞ = characteristic impedance of a free pile
γ = transfer constant
r_0 = pile radius
R_L = pile stiffness
r_L = soil stiffness
φ, ψ = horizontal and vertical wave functions
M_L = material mass
R_e = equivalent stiffness of the pile at low frequency

In the following, the letters corresponding to phenomena exterior to the pile are 'primed'.

INTERNAL PILE-WAVE PROPAGATION

The methods and notations used by Press, Ewing and Jardesky in *Elastic Waves in Layered Media* (Press *et al.*, 1957) are followed here. This amounts to searching for combinations of wave functions with separation of variables that satisfy extreme conditions. The functions to be used in cylindrical coordinates are of the form:

$$\varphi = AF(r)e^{i(\omega t - \upsilon z)}, \text{ for longitudinal waves}$$
$$\psi = CG(r)e^{i(\omega t - \upsilon z)}, \text{ for transverse waves}$$

and γ(the number of waves along the z-axis) is taken as being equivalent to the limiting conditions along the z-axis, whatever they may be.

The parameters φ and ψ must satisfy the relations:

$$\nabla^2\varphi = (1/\alpha^2)(\partial\varphi^2/\partial t^2), \nabla^2\psi = (1/\beta^2)(\partial\psi^2/\partial t^2) \quad \text{(AI.1)}$$

where α and β are the velocities of pure longitudinal and transverse waves.

When the functions φ and ψ are defined, the displacements w and q are derived from the following relations:

$$w = (\partial\varphi/\partial z) - (\partial^2\psi/\partial r^2) - (1/r)(\partial\psi/\partial r), q = (\partial\phi/\partial r) + [\partial^2\psi/(\partial r\partial z)] \quad \text{(AI.2)}$$

The stresses, $p_{rr} p_{rz}$ and p_{zz}, are obtained from:

$$p_{rr} = \lambda(q/r) + (\partial q/\partial r) + (\partial w/\partial z) + 2\mu(\partial q/\partial r) \quad \text{(AI.3a)}$$
$$p_{rz} = \mu(\partial q/\partial z + \partial w/\partial r) \quad \text{(AI.3b)}$$
$$p_{zz} = \lambda(q/r + \partial q/\partial r + 3\partial w/\partial z) \quad \text{(AI.3c)}$$

From equation (AI.1):

$$\partial^2 F/\partial r^2 + (1/r)(\partial F/\partial r) + (k_\alpha^2 - v^2)F(r) = 0, \text{ where } k_\alpha = \omega/\alpha \quad \text{(AI.4a)}$$
$$\partial^2 G/\partial r^2 + (1/r)(\partial G/\partial r) + (k_\beta^2 - v^2)G(r) = 0, \text{ where } k_\beta = \omega/\beta \quad \text{(AI.4b)}$$

Let $k_\alpha^2 - v^2 = k^2$ and $k_\beta^2 - v^2 = k_1^2$. F and G are then of the form:

$$aJ_0(kr) + bY_0(kr) \quad \text{(AI.5)}$$

where J_0 is the Bessel function of the first kind (order zero) and Y_0 is the Bessel function of the second kind (order zero).

Simplifying the basic hypothesis, the following assumptions can be made:

$$\lambda = \mu \text{ and } w = (-\mathrm{i}vA + Ck_i^2)\mathrm{e}^{-\mathrm{i}vz} \quad \text{(AI.6a)}$$
$$q = (-Ak^2/2 + \mathrm{i}vCk_1^2/2)r\mathrm{e}^{-\mathrm{i}vz} \quad \text{(AI.6b)}$$

where A is homogenous to L^2 and C is homogeneous to L^3. Furthermore:

$$p_{rr}/\lambda = A\mathrm{i}vk^2r - C(k_1^2/2)r(k_\beta^2 - 2v^2) \quad \text{(AI.7)}$$

For a free cylinder, $p_{rr} = 0$ and $p_{rz} = 0$, which gives the relation:

$$(k^2 + k_\alpha^2)k_1^2/2(k_\beta^2 - 2v^2) + v^2k^2k_1^2 = 0 \quad \text{or} \quad (k^2 + k_\alpha^2)(k_\beta^2 - 2v^2) + 2v^2k^2 = 0 \quad \text{(AI.8)}$$

This gives: $v^2 = 6/5k_\alpha^2$ and $C = \sqrt{(5/6)}\alpha = 0.915\alpha$.

For $\alpha = 4000$ m s^{-1} and $c = 3700$ m s^{-1}, where the bar-wave velocity is defined by $c = \sqrt{E/\rho}$, the displacements w and q can then be calculated as follows:

$$\text{for } p_{rr} = 0,\ A = \mathrm{i}C9/4\sqrt{6/5}k_\alpha$$
$$w = W\mathrm{e}^{-\mathrm{i}vz} \quad \text{(AI.9a)}$$
$$q = 0.2Wk_\alpha r\mathrm{e}^{-\mathrm{i}vz} \quad \text{(AI.9b)}$$

where W is the amplitude of the displacement w.

As previously discussed, $k_\alpha r$ is much less than 1 and the radial displacement q is negligible compared to the axial displacement.

The velocity along the x-axis is given by:

$$\partial w/\partial t = \mathrm{i}\omega W\mathrm{e}^{-\mathrm{i}vz} \quad \text{(AI.10)}$$

CALCULATION OF ENERGY TRANSMITTED TO THE PILE

The stress p_{zz} can be calculated as:

$$p_{zz} = \lambda(q/r + \partial q/\partial r + 3\partial w/\partial z) \approx 3\lambda(\partial w/\partial z) \tag{AI.11}$$

and from equation (AI.9):

$$p_{zz} = -3\mathrm{i}v\lambda W \mathrm{e}^{-\mathrm{i}vz} \tag{AI.12}$$

The transmitted energy per unit section area is:

$$P_{\mathrm{u}} = p_{zz}/2(\partial w/\partial t) = 3/2\lambda v\omega W^2 \tag{AI.13}$$

while the total transmitted energy is:

$$P_{\mathrm{t}} = 3/2\pi\lambda v\omega r^2 W^2 = 3/2\pi(\lambda r^2/c)\omega^2 W^2 \tag{AI.14}$$

The transported energy is proportional to the square of the velocity ωW and to the cross-sectional area πr^2.

RADIATED ENERGY: WAVES IN THE SOIL

As in all cylindrical bodies including wave propagation to infinity, the solution to equation (AI.5) requires a Hankel function, defined by:

$$H_0(kr) = J_0(kr) - jY_0(kr) \tag{AI.15}$$

The asymptotic form of the above is:

$$H_0(kr) = \sqrt{(2/\pi kr)}\mathrm{e}^{-j(kr-\pi/4)} \tag{AI.16}$$

The functions ϕ' and ψ' relative to the external medium (the soil) are:

$$\varphi' = A' H_0(k'r)\mathrm{e}^{[\mathrm{i}(\omega t - vz)]} \tag{AI.17a}$$

$$\psi' = C' H_0(k'r)\mathrm{e}^{[\mathrm{i}(\omega t - vz)]} \tag{AI.17b}$$

In order to determine the coefficients A' and C', it can be assumed that for $r = r_0$ (pile radius) the displacements are identical in both pile and soil. It can also be assumed that $k'r$ and $k'_1 r$ are small in all that follows below.

The function H_0 can then be written as:

$$H_0(kr) = -2/\pi[\gamma - \log(kr/2)]$$
$$\varphi' = A'2/\pi[\gamma - \log(k'r/2)]e^{-ivz}$$
$$\psi' = 2C'/\pi[\gamma - \log(k_1'r/2)]e^{-ivz}$$
$$q' = \partial\varphi'/\partial r + \partial^2\psi'/\partial r\partial z = 2A'/\pi r - iv2C'/\pi r = 2/\pi r(A' - ivC') \tag{AI.18a}$$

$$w' = \partial\varphi'/\partial z + \partial^2\psi'/\partial r - (1/r\partial\psi'/\partial r) = \partial\varphi'/\partial z + k_1'\psi'$$
$$= -2ivA'/\pi(\gamma - \log k'r/2) + 2C'/k_1'^2(\gamma - \log k_1'r/2) \tag{AI.18b}$$

The identity of the displacements gives:

$$q = q'; 0.2Wk_\alpha r_0 = 2/\pi r_0(A' - ivC') \tag{AI.19a}$$
$$w = w'; W = -2ivA'/\pi(\gamma - \log k'r_0/2) + 2C'k_1'^2/\pi(\gamma - \log k_1'r_0/2) \tag{AI.19b}$$

When the approximation $kr \ll 1$:

$$A' = ivWr/[2k'^2(\gamma - \log k'r_0/2)] \text{ and } C' = \pi W/[2k_1'^2(\gamma - \log k_1'r_0/2)] \tag{AI.20}$$

The asymptotic form then becomes:

$$\varphi' = ivW\pi/[2k'^2(\gamma - \log k'r_0/2)]\sqrt{(2/\pi kr)}e^{i(-k'r - vz + \omega t)}e^{i4/\pi} \tag{AI.21}$$

The wave surfaces have the cone equation, $k'r - vz = 0$, and then:

$$\tan\gamma = v/k' \text{ and } \tan\gamma_1 = v/k_1' \tag{AI.22a}$$
$$\tan\alpha = v/\sqrt{(k_{\alpha'}^2 - v^2)} \text{ and } \sin\alpha = v/k_{\alpha'} = \alpha'/C \tag{AI.22b}$$

2. DETERMINATION OF DAMPING

$$\boldsymbol{a} = \text{(radiated energy per meter)/(transmitted energy)}^2$$
$$\boldsymbol{a} = P_r'/P_t = \left\{W^2\pi^2\mu'\omega c/\left[2\left(\gamma - \log k_1'r_0/2\right)^2\right]3/2\pi\lambda r^2\omega^2W^2\right\}$$
$$= \pi/3\mu'/\gamma\left\{c/\omega r^2\left[\left(\gamma - \log k_1'r_0/2\right)^2\right]\right\} \tag{AI.23}$$

As a numerical example for:

$$\mu'/\gamma = 1/20, c = 3700\,\text{m s}^{-1} \text{ and } r_0 = 0.3\,\text{m}$$
$$a = 1/20 \times 3700/(0.09 \times 16)1/\omega = 110/\omega$$

and so for $\omega = 600$, $\boldsymbol{a} = 0.18\ \text{Np m}^{-1}$.

[2] In neper (Np) per meter.

This damping is significant, since for a 10-m long pile the calculated damping for a 'round-trip' stress-wave signal is 3.6 Np (31 dB). The damping coefficient, ***a***, decreases asymptotically with increasing frequency. This expression shows that damping is inversely proportional to the square of the pile radius. Therefore, for piles with diameters of 250 mm or less, the impedance method has very little chance of success, except for very short shafts.

EVALUATION OF THE ELASTIC SOIL REACTION (LATERAL ANCHORAGE) FOR AN INFINITELY LONG PILE

The vertical displacement, w', of the soil is given by:

$$w' = -2\mathrm{i}vA'(\gamma - \log k'r/2) + 2C'k_1'^2/\pi(\gamma - \log k_1'r/2) \qquad \text{(AI.24a)}$$

$$q' = 2/\pi r(A' - \mathrm{i}vC') \qquad \text{(AI.24b)}$$

Replacing A' and C' by their values:

$$w' = W(\gamma - \log k_1'r/2)/(\gamma - \log k_1'r_0/2)$$

The stress, p_{rz}, is then given by:

$$\mu'(\partial q'/\partial z + \partial w'\partial r) = \mu'\partial w'/\partial r \text{ or } p_{rz} = \mu'W/r_0(\gamma - \log k'r_0/2) \qquad \text{(AI.25)}$$

The force per unit length (assumed small with respect to the wavelength) is:

$$F_\mathrm{u} = 2\pi r_0 p_{rz} = 2\pi\mu'W/(\gamma - \log k'r_0/2) \qquad \text{(AI.26)}$$

while the unit stiffness is $2\pi\mu'/(\gamma - \log k'r_0/2)$.

This unit reaction allows the calculation of pile behavior at low frequencies.

3. DETERMINATION OF HARMONIC RESPONSE – MECHANICAL IMPEDANCE

A vertical sinusoidal force applied to a pile head produces a resultant pile head velocity. The ratio of the complex amplitudes F and v is known as the 'Mechanical Impedance' of the pile/soil system. The impedance is a function of the pulse phase, i.e. $Z(j\omega) = F/v$.

The impedance can be measured experimentally, by either a swept-frequency vibrator or by impulse hammer, with a velocity transducer. Only the modulus of the impedance Z is measured.

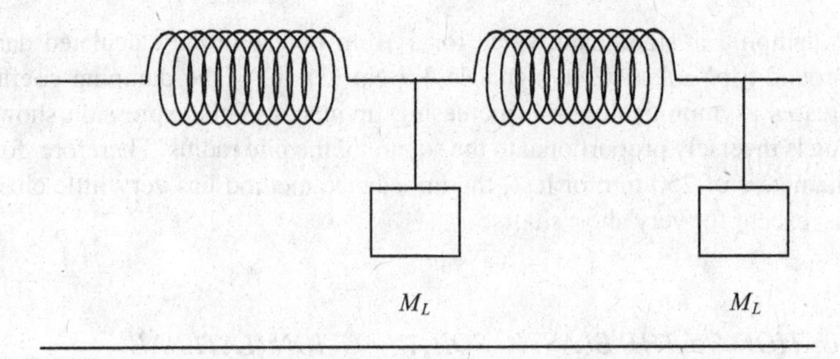

Figure AI.1 Electrical analog of an infinitely long free pile

CALCULATION METHOD – GENERAL CASE

The pile/soil system is reduced to a simple line transmission system. The characteristic parameters for this system are used, i.e. the characteristic impedance, Z_c, and the transmission constant, γ. The entry impedance for this system, assumed closed at its base, is then calculated for a given mechanical impedance. An initial assumption is the calculation of the characteristic parameters are made for an infinitely long pile, which are then used for pile/soil systems of finite depths.

An infinitely long free pile can be likened to a thin bar or rod, and in this case, $Z_{c0} = \rho c S$ and $\gamma_0 = j\omega/c$.

These values can be obtained from linear impedance theory, both in series and in parallel, from the relations:

$$Z_{c0} = \sqrt{(Z_s Z_p)} \quad \text{and} \quad \gamma_0 = \sqrt{(Z_p/Z_s)}$$

For the free pile (Figure AI.1):

$$Z_s = R_L/j\omega \quad \text{and} \quad Z_p = j\omega M_L \qquad \text{(AI.27a)}$$

$$Z_{c0} = \sqrt{(R_L M_L)} \quad \text{and} \quad \gamma_0 = \sqrt{[jM_L\omega(j\omega/R_L)]} = j\omega\sqrt{(M_L/R_L)} \qquad \text{(AI.27b)}$$

If $R_L = SE$ and $M_L = \rho S$, then:

$$Z_{c0} = \sqrt{(E\rho S^2)} = \sqrt{(E/\rho)}\rho S = \rho c S \quad \text{and} \quad \gamma_0 = j\omega\sqrt{(\rho S/SE)} = j\omega/c \qquad \text{(AI.28)}$$

For an infinite pile buried in soil, the elementary analogy is modified as shown in Figure AI.2. The soil is represented by a spring and a dashpot, with coefficients r_L and $\boldsymbol{a}$, respectively. These coefficients are determined at any frequency by the relationships presented in the preceding chapters on wave propagation in cylindrical structures, and in particular:

$$r_L = (2\pi\mu')/(\gamma - \log k' r_0/2)$$

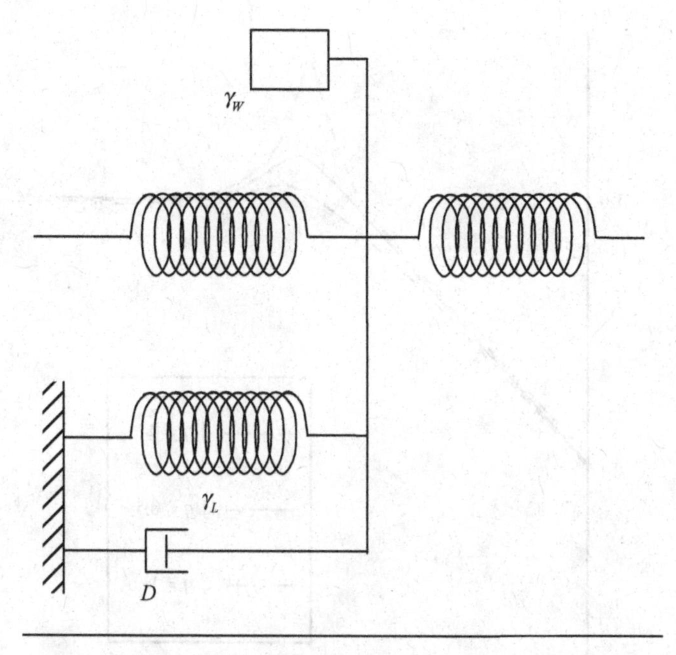

Figure AI.2 Electrical analog of a pile in soil

Z_s and Z_p are series and parallel resistances, respectively:

$$Z_s = R_L/j\omega \text{ and } Z_p = M_L p + \boldsymbol{a} + r_L/p, \text{ where } j\omega = p$$

$$Z_c = \sqrt{(Z_s Z_p)} = \sqrt{[R_L/p(M_L p + \boldsymbol{a} + r_L/p)]}$$

$$= \sqrt{(R_L M_L)}\sqrt{[(p^2 + \boldsymbol{a}p/M_L + r_L/M_L)/p^2)]}$$

$$Z_c = Z_{c0}\sqrt{[(p^2 + \xi p + \eta^2)/p^2]}, \text{ where } \xi = \boldsymbol{a}/M_L \text{ and } \eta^2 = r_L/M_L \quad \text{(AI.29)}$$

In the above, ξ and η are homogeneous with the pulsation. It can be seen that as $p \to \infty$, $Z_c \to Z_{c0}$.

Figure AI.3 gives the amplitude of the phase of $Y_c = 1/Z_c$ as a function of frequency for different values of the damping ratio, η/ξ. For low frequencies, $Z_c \to \sqrt{(R_L r_L/p)}$.

Since the characteristic impedance is also equal to the resistance of the entry impedance for an infinitely long pile, it can be seen that at low frequencies, the pile is equivalent to a spring whose stiffness is a function of both the pile and soil characteristics. However, it should be noted that the stiffness, r, is not constant.

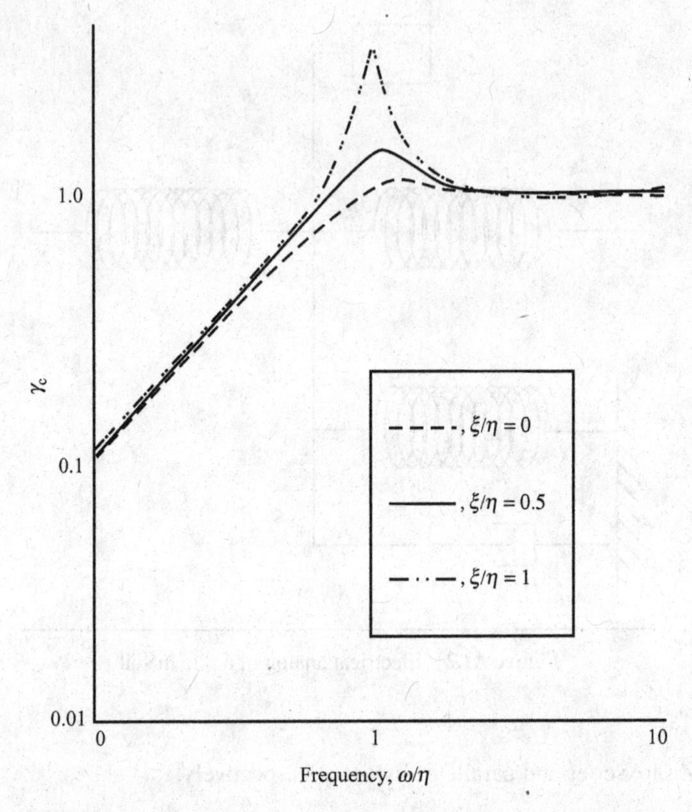

Figure AI.3 The effect of damping as a function of frequency

4. RESONANT FREQUENCY OF AN INFINITELY LONG PILE

Y_c includes a maximum for the pulsation, η, if ξ/η is small. This comprises a mass resonance for the pile and its lateral reaction. For a very long pile, this resonance is independent of its length.

The propagation constant, γ, is given by:

$$\gamma = \sqrt{(Z_p/Z_s)} = \sqrt{[(M_L p + \mathbf{a} + r_L/p)p/R_L]}$$
$$= \sqrt{(M_L/R_L)}\sqrt{(p^2 + \boldsymbol{a}p/M_L + r_L/M_L)} \qquad \text{(AI.30a)}$$

$$\gamma = \gamma_0\sqrt{[(p^2 + \xi p + \eta^2)/p]} \qquad \text{(AI.30b)}$$

Figure AI.4 illustrates the variations in the real and imaginary parts of γ as a function of ω with various constant values of the parameters ξ and η.

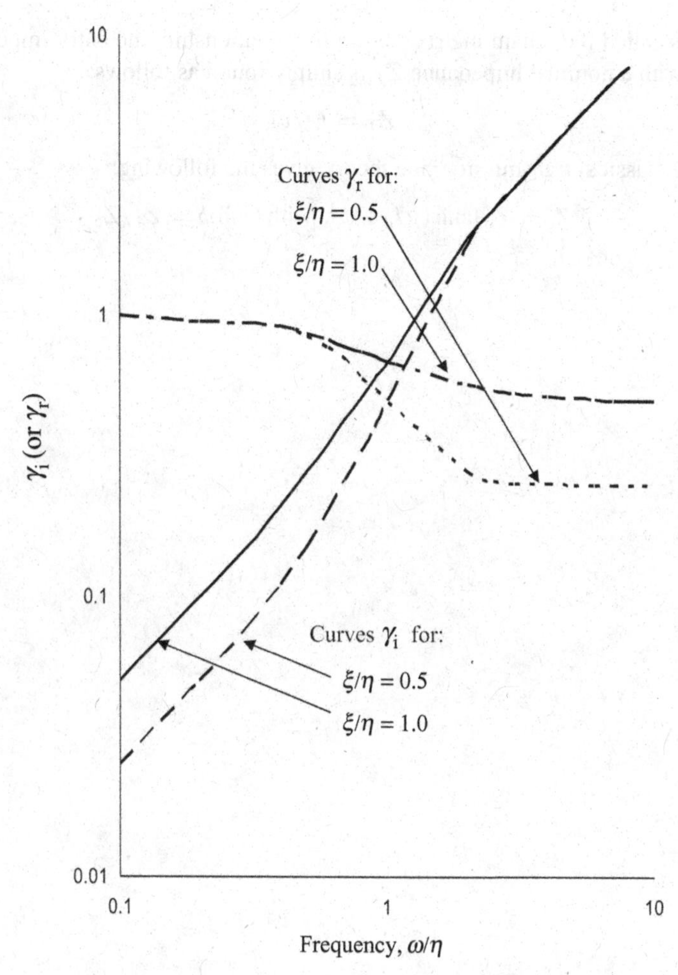

Figure AI.4 Variations in the real and imaginary parts of y

5. IMPEDANCE INPUT FOR A FINITE LENGTH PILE WITH UNKNOWN MECHANICAL IMPEDANCE AT ITS BASE

Imagine a mechanical transmission line with length L, characteristic impedance Z_c and propagation constant γ. F_1 and v_1 are the entry force and velocity, while F_2 and v_2 are the exit force and velocity, respectively. The forces and velocities are related by the following:

$$F_1(\cosh \gamma L \times Z_c \sinh \gamma L)F_2 = v_1[\sinh \gamma L / Z_c \cosh \gamma L)v_2 \qquad \text{(AI.31)}$$

This is called the 'chain matrix'. From this relationship, the entry impedance for the pile with a nominal impedance Z_T is simply found as follows:

$$Z_T = F_2/v_2$$

while the classical transmission line theory gives the following:

$$Z_e = Z_c \tanh(\gamma L + \phi), \text{ with } \tanh\phi = Z_T/Z_c \tag{AI.32}$$

Appendix II Contact Addresses

For the convenience of the reader who may want to seek further information or obtain copies of some of the research references, standards or codes of practice mentioned in this book, the addresses of the key research, standardization and/or sponsoring organizations are given below.

ACI

(American Concrete Institute)

ACI International	Phone: +1 248-848-3700
38800 Country Club Drive	Fax: +1 248-848-3701
Farmington Hills, MI 48331	www.aci-int.org
USA	

ADSC-IAFD

(The International Association of Foundation Drilling – formerly the Association of Drilled Shaft Contractors)

ADSC-IAFD	Phone: +1 214-343-2091
PO Box 550399	Fax: +1 214-343-2384
Dallas, TX 75355-0339	www.adsc-iafd.com
USA	

AFNOR

(Association Française de Normalisation)

11, Avenue Francis de Pressensé
93571, Saint-Denis la Plaine
Codex, France

Phone: +33 (0)1 41 62 80 00
Fax: +33 (0)1 49 17 90 00
www.afnor.fr

ASCE

(American Society of Civil Engineers)

American Society of Civil Engineers
1801 Alexander Bell Drive
Reston, VA 20191
USA

Phone: +1 703-295-6300
Fax: +1 703-295-6222
www.asce.org

ASTM

(American Society for Testing and Materials)

ASTM International
PO Box C700
100 Barr Harbor Drive
West Conshohocken, PA 19428–2959
USA

Phone: +1 610-832-9585
Fax: +1 610-832-9555
www.astm.org

BSI

(British Standards Institute)

BSI Global
389 Chiswick High Road
London W4 4AL
United Kingdom

Phone: +44 (0)20 8996 9000
Fax: +44 (0)20 8996 7001
www.bsi-global.com

CEBTP

(Centre Experimentale de Recherche et d'Études du Bâtiment et des Travaux Publics)

CEBTP
Domaine de Saint-Paul – BP 37
78470, St. Remy Les Chevreuse
France

Phone: +33 (0)1 30 85 24 00
Fax: +33 (0)1 30 85 24 30
www.cebtp.fr

CIRIA

(The Construction Industry Research and Information Association)

CIRIA	Phone: +44 (0)20 7222 8891
6 Storey's Gate	Fax: +44 (0)20 7222 1708
London SW1P 3AU	www.ciria.org.uk
United Kingdom	

DFI

(The Deep Foundations Institute)

DFI International	Phone: +1 973-423-4030
326 Lafayette Avenue	Fax: +1 973-423-4031
Hawthorne, NJ 07506	www.dfi.org
USA	

DGGT

(German Society for Geotechnique)

Deutsche Gesellschaft für Geotechnik	Phone: +49 (02) 01 782723
Hohenzollernstrasse 52	Fax: +49 (02) 01 782743
D-45128 Essen	www.dggt.de
Germany	

DIN

(Deutches Institüt fur Normung)

DIN Deutsches Institut für Normung eV	Phone: +49 (30) 2601-0
Burggrafenstrasse 6	Fax: +49 (30) 2601-1231
10787 Berlin	www.din.de
Germany	

FHWA

(Federal Highway Administration)

FHwA East Coast Resource Center	Phone: +1 410-962-0093
10 South Howard Street, Suite 400	Fax: +1 410-962-3655
Baltimore, MD 21201-2819	www.fhwa.dot.gov
USA	

Note: The FHwA has a large number of field offices throughout the USA. For US enquiries, go to the Web Site and look up the field office nearest you. For overseas enquiries, we have listed the East Coast Resource Center – hopefully they will be able to answer your queries or direct you to the appropriate office.

ICE

(Institution of Civil Engineers)

ICE
1 Great George Street
London SW1P 3AA
United Kingdom

Phone: +44 0(20) 7222 7722
www.ice.org

ISSMGE

(International Society for Soil Mechanics and Geotechnical Engineering)

ISSMGE Secretariat
Geotechnical Engineering Research Centre
City University
Northampton Square
London EC1V 0HB
United Kingdom

Phone: +44 0(20) 7040 8154
Fax: +44 0(20) 7040 8832
www.issmge.org

STANDARDS AUSTRALIA

(Standards Association of Australia)

Standards Australia
GPO Box 5420
Sydney NSW 2001
Australia

Phone: 1300 65 4646 (within Australia)
+61(2) 8206 6010 (International)
Fax: 1300 65 4949 (within Australia)
+61(2) 8206 6020 (International)
sales@standards.com.au

TNO

(The Netherlands Organization for Building and Construction Research)

TNO-Bouw
Postbus 49
NL-2600 AA Delft
The Netherlands

Phone: +31 (15) 276 3000
Fax: +31 (15) 276 3010
www.bouw.tno.nl

Appendix III Standards Referred to in this Book

For the reader's convenience, the national and international standards that have been developed for the test methods described in this book have been 'cross-referenced'. Most standards can be purchased and downloaded direct from the respective Standards Organizations' websites. All European standards also appear to be available on the AFNOR website at www.afnor.fr. This website can be accessed in both English or French.

Please note that 'hunting' for foreign standards using only generic keywords is a time-consuming business. This list contains only those standards that could be located in a reasonable amount of time. It should not, therefore, be considered 'complete'. If an applicable standard for your home state or country is not listed here, please don't take it personally! If you have time, please let the present authors know of their omission by contacting Bernie Hertlein at hertlein@stsconsultants.com – who knows? – maybe we'll be invited to revise this book in another ten years time!

1. CROSS-HOLE SONIC LOGGING

AFNOR: 'Sols – Reconnaissance et essais: Auscultation d'un élément de fondation, Partie 1 – Méthode par transparence [(Soils: Investigation and Testing: Testing a Foundation Element, Part 1 – Sonic Coring Method (Cross-Hole Sonic Logging)]'.

ASTM D6760: 'Standard Test Method for Integrity Testing of Deep Foundations by Cross-Hole Testing'.

Nondestructive Testing of Deep Foundations B.H. Hertlein and A.G. Davis

ICE: 'Specification for Piling and Embedded Retaining Walls: Specification, Contract Documentation and Measurement: Guidance Notes'.

2. GAMMA–GAMMA LOGGING

AFNOR XP P94-160-5: 'Sols – Reconnaissance et essais: Auscultation d'un élément de fondation, Partie 5 – Méthodes diffusion nucleaire à rayonnement gamma [Soils: Investigation and Testing: Testing a Foundation Element, Part 5 – Diffusion of Gamma Rays Method (Gamma–Gamma Logging)]'.

3. HIGH-STRAIN TESTING OF PILES

AFNOR XP P94-152: 'Sols – Reconnaissance et essais: Essai de chargement dynamique axial d'un élément de fondation profonde [Soils: Investigation and Testing: Dynamic Axial Load Test of a Deep Foundation Element]'.

ASTM D4945: 'Standard Test Method for Dynamic Testing of Piles'.

Australian Standard AS2159-1995: 'Piling – Design and Installation'.

CAN/CSA S6-00: 'Canadian Bridge Design Code (formerly Ontario Highway Bridge Design Code)'.

ICE: 'Specification for Piling and Embedded Retaining Walls: Specification, Contract Documentation and Measurement: Guidance Notes'.

4. IMPULSE-ECHO AND IMPULSE-RESPONSE TESTS

AFNOR NF P94-160-2: 'Sols: Reconnaissance et essais: Auscultation d'un élément de fondation, Partie 2 – Méthode par réflexion [Soils: Investigation and Testing: Testing a Foundation Element, Part 2 – Reflection Method (Impulse Echo)]'.

AFNOR NF P94-160-4: 'Sols – Reconnaissance et essais: Auscultation d'un élément de fondation, Partie 4 – Méthode par impédance [Soils: Investigation and Testing: Testing a Foundation Element, Part 4 – Impedance Method (Impulse Response)]'.

ASTM D5882: 'Standard Test Method for Low-Strain Integrity Testing of Piles'.

ICE: 'Specification for Piling and Embedded Retaining Walls: Specification, Contract Documentation and Measurement: Guidance Notes'.

5. PARALLEL SEISMIC

AFNOR NF P94-160-3: 'Sols – Reconnaissance et essais: Auscultation d'un élément de fondation, Partie 3 – Méthode sismique paralléle [Soil: Investigation and Testing: Testing a Foundation Element, Part 3 – Parallel Seismic Method]'.

6. STATIC LOAD TESTING OF DEEP FOUNDATION SHAFTS

AFNOR NF P94-150-1: 'Sols – Reconnaissance et essais: Essai statique de pieu isolé sous un effort axial, Partie 1 – en compression [Soils: Investigation and Testing: Static Test of a Single Pile Under Axial Load, Part 1 – In Compression]'.

AFNOR NF P94-150-2: 'Sols – Reconnaissance et essais: Essai statique de pieu isolé sous effort axial, Partie 2 – en traction [Soils: Investigation and Testing: Static Test of a Single Pile under Axial Load, Part 2 – In Tension]'.

AFNOR NF P94-151: 'Sols – Reconnaissance et essais: Essai statique de pieu isolé sous effort transversal [Soils: Investigation and Testing: Static Test of a Single Pile under Lateral Load]'.

ASTM D1143: 'Standard Test Method for Piles under Static Axial Compressive Load'.

ASTM D3689: 'Standard Test Method for Individual Piles under Static Tensile Load'.

ASTM D3966: 'Standard Test Method for Piles under Lateral Loads'.

BS 8004-2: 'Constant Rate of Penetration Test'.

DGGT: 'Empfehlungen des Arbeitskreises 2.1 der DGGT für statische und dynamische Pfahlprüfungen [Recommendations of Committee 2.1 of the German Society for Geotechniques for Static and Dynamic Pile Testing]'.

ICE: 'Specification for Piling and Embedded Retaining Walls: Specification, Contract Documentation and Measurement: Guidance Notes'.

Appendix IV Sample Specifications for NDT Methods for Deep Foundations

Differing contracting practices around the world make it impossible to come up with one universal specification for any of the deep foundation tests. Depending on who actually hires the testing firm, whether it be the owner, the engineer or the contractor, the specification may be a stand-alone document, or it may be rolled into a larger specification for the foundation construction or some other part of the project. Possibly the best example in the USA of a combined specification for both construction and testing with the CSL method is that which was developed by Washington State Department of Transportation (DOT) with the help of the West Coast Chapter of ADSC.

Other useful sample specifications have been promulgated by various manufacturers of testing equipment, but, perhaps not surprisingly, some of those specifications tend to favor a particular manufacturer's equipment or data presentation format.

The following sample specifications are 'composites' that encompass what the present authors consider to be some of the better sample specifications currently available on the Internet, tempered with our own experiences. These specifications include a number of optional paragraphs and typical quantities in parentheses to guide the user in tailoring the specification to suit the specific project requirements. Acknowledgement is made to Washington State DOT and equipment manufacturers Olson Instruments, Pile Dynamics Inc. and Testconsult, for the various test specifications that they wrote, extracts of which were used in compiling the following samples.

Nondestructive Testing of Deep Foundations B.H. Hertlein and A.G. Davis

1. SAMPLE SPECIFICATION FOR LOW-STRAIN TESTING BY EITHER IMPULSE ECHO OR IMPULSE RESPONSE

Note: Any quantities given in braces {*nn*} in the following specifications are only recommendations. The actual numbers may be varied to suit project requirements and/or site-specific conditions.

OVERVIEW OF LOW-STRAIN TESTING

This specification is concerned with low-strain integrity tests for deep foundations. These integrity tests are performed with a small hand-held hammer, as opposed to the methods that require the use of a drop-weight, pile-driving hammer or other means for generating a 'high-strain' wave in the shaft. The two basic forms of the low-strain method are the Impulse-Echo (or Sonic-Echo) method, in which data is analyzed in the time domain, or the Impulse-Response (Sonic-Mobility) method, in which data is analyzed in the frequency domain. Both methods are described in, and shall be performed in accordance with ASTM D5882, 'Standard Test Method for Low Strain Integrity Testing of Piles'.

The low-strain integrity test methods are applicable to concrete and timber piles, drilled shafts, and augered, cast-in-place piles. They are generally unsuitable for steel H-piles and sheet-piles, unless the piles are relatively short or installed in soft soils. There are several proprietary versions of the methods, any of which may meet the intentions of this specification provided that the specified information is obtained and the specified data-presentation format is used in reporting test results.

TEST METHOD DESCRIPTION

The low-strain integrity test requires the generation of an elastic wave by striking the top of the shaft with a hand-held hammer. For the Impulse-Echo method the hammer is usually not instrumented and the shaft top response is measured by an acceleration transducer (accelerometer). For the Impulse-Response method, the hammer contains a force transducer to measure the force input to the shaft and the response of the shaft is usually measured with a velocity transducer (calibrated geophone). The frequency range of the energy input by the hammer impulse varies according to the shape and size of the hammer, and the material of the hammer impact face. A hard rubber or nylon face on a #2 sledge hammer has been found to be appropriate for most shaft/soil geometries, but an aluminum face may be used to help resolve details in the upper portion of the foundation.

The elastic compression wave generated by the hammer impulse propagates down the foundation shaft at a velocity that is a function of the modulus and density of the material. Changes in the dimensions or material of the shaft, and significant changes in soil conditions will reflect all or part of the wave energy back to the head of the shaft. The depth of the reflector can be calculated from the travel time of the reflected wave if the wave velocity is assumed. Conversely, if the depth is known, the velocity can be calculated.

The frequency range of the response detected by the motion transducer will depend on the coupling between the transducer and the surface of the shaft. The more rigid the connection, the broader will be the frequency range of the response. In some cases, it may be possible to simply hold the transducer in contact with the test surface, but more usually some type of couplant is used to provide a good connection. Suitable couplants range from adhesives or putties to stiff grease or gel.

Impulse Echo (Sonic Echo)

For the Impulse-Echo method, the shaft-head response is displayed in terms of either velocity or acceleration as a function of time. Several measurements shall be recorded to verify signal consistency, and the results averaged. The averaged velocity trace is generally regarded as the standard result of the Impulse-Echo method.

Where wave attenuation caused by stiff soils or a high length/diameter ratio result in weak signals, an exponential amplification function may be applied to enhance reflections from the lower portion of the foundation.

Impulse Response (Sonic Mobility)

For the Impulse-Response method the time-based force and velocity data are converted into the frequency domain by Fast Fourier Transform (FFT), where velocity is divided by force to provide the transfer function, or characteristic mobility plot. The test result is presented as a graph of mobility as a function of frequency. The low-frequency portion of the graph provides a measure of the dynamic stiffness of the shaft/soil complex. As with the Impulse-Echo method, several measurements shall be recorded to verify data consistency, and averaged to provide the final result. The mobility curve, with dynamic stiffness measurement, is generally regarded as the standard result of the Impulse-Response method.

Complex responses caused by multiple reflectors or signals attenuated by soil damping may be enhanced with the Impedance-Log analysis. For this procedure, the time-domain data are amplified in a similar manner to the Impulse-Echo data, and processed to generate a 'Reflectogram', which identifies the main reflections. A computer program is then used to generate a simulated response. The parameters for the simulation are modified until a good match with the test data is achieved. The software then generates a graphic representing the effective impedance over the length of the shaft.

TEST EQUIPMENT REQUIREMENTS

The data-acquisition equipment shall be digital, with a minimum resolution of 12 bits. The data-acquisition equipment, plus any signal conditioning or power supply devices used in the testing process must have a signal-to-noise ratio appropriate for the accurate acquisition of low-amplitude signals under field conditions. The sampling frequency (analog-to-digital or A/D conversion rate) shall be at least 25 kHz.

For the Impulse-Echo method, a single-channel data-acquisition system may be used. For the Impulse-Response method, the hammer shall contain a transducer to measure the force input, and the data-acquisition system shall have at least two channels, in order to record force and velocity data simultaneously. The means of coupling the motion transducer to the surface of the shaft shall be sufficiently rigid to ensure undistorted response over the frequency range of interest.

Data shall be displayed in the field to allow verification of data quality and preliminary interpretation. The data shall be stored in a form that permits subsequent additional processing or analysis.

QUALIFICATIONS OF PERSONNEL

The field testing, and preliminary interpretation if required, shall be performed by an experienced technician with at least {one (1)} year's experience in integrity testing with the particular method being applied. Final interpretation and reporting shall be performed by, or under the direct supervision of, an engineer or senior technician with at least {three (3)} years' experience in integrity testing with the specific method being applied.

Experience in the application of integrity test methods other than the method being employed for this project, and/or experience in analysis of test data gathered by those methods, will be deemed irrelevant for the purposes of this specification.

SHAFT HEAD PREPARATION

For shafts where concrete or grout was cast in place, integrity testing shall not be performed until the concrete has cured for a minimum of {five (5)} days, unless otherwise directed or approved by the engineer of record.

In all cases, the heads of the shafts to be tested shall be free of water, soil and site debris. The ground around the head of the shaft shall be excavated or graded such that the shaft head remains clear of any standing water. The concrete or grout of the shaft head shall be sound and undamaged. The center of the shaft head and at least one location near the perimeter of the shaft shall be prepared for application of the test transducers. Any laitance or other contamination must be removed with either a grinder, a scabbling tool or a bush hammer. The use of jackhammers to prepare the test locations shall not be permitted. In the event that jackhammers are used to cut back the head of the shaft before the integrity testing is performed, then the test transducer locations shall be prepared with a either a grinder, a scabbling tool or a bush hammer to remove any cracked concrete. Each test transducer location shall be a flat and level area of sound concrete or grout at least 50 mm (2 in) in diameter. As a rule-of-thumb guide, if a small soda bottle can stand upright without rocking, the area is smooth enough.

At least {25} % of the shafts shall be integrity tested. Selection of the shafts to be tested will be performed by the engineer of record. Where less than 100 % of the shafts are tested, additional shafts may be selected for testing at the discretion of the engineer if unusual circumstances are noted during construction, or if damage is suspected to have occurred as a result of other construction activity, such as equipment movement or excavation.

REPORTING TEST RESULTS

The testing organization shall present a written report within {five (5)} working days after performing the tests. Geotechnical exploration borehole logs and foundation drilling records shall be made available to the testing organization upon request. The report shall include, for each shaft tested, the averaged velocity versus time record if the Impulse-Echo test was performed, or the mobility response curve if the Impulse-Response test was performed. The report shall also contain a table that summarizes the test results and interpretation. The text of the report shall include a paragraph of conclusions that summarizes the opinion of the testing organization as to the integrity of the shafts tested. Additional plots and graphics may be included, as required by the engineer or recommended by the testing organization.

SHAFT ACCEPTANCE OR REJECTION CRITERIA

If a clear response from the shaft toe is detected in the results of either Impulse-Echo or Impulse-Response testing, with no significant secondary responses from reflectors located above the shaft toe, the shaft shall be deemed acceptable. If no clear response from the shaft toe is detected, and no significant secondary reflections are observed, the senior engineer with the testing organization shall confer with the engineer of record for the project to determine what length of the foundation may be considered proven, based on the inconclusive test data.

Reflections interpreted to be from any level above the level of the toe of the shaft shall be analyzed to determine if the reflection is from a significant reduction or increase in cross-section. The probable nature and effect of the reflector shall then be assessed, considering the intent behind the design of the shaft. The decision to accept or reject the shaft shall be made by the engineer of record, after consultation with an experienced engineer from the testing organization and/or the geotechnical engineer for the project.

In the event that the test result is deemed to be inconclusive, trimming back the pile head and retesting may provide more conclusive data. If this measure is impractical or does not succeed, additional testing by other means, including, but not limited to, core sampling and/or Sonic-Log testing, high-strain testing or full-scale load testing, may be required.

BASIS FOR PAYMENT

Payment for Low-strain Integrity Testing

The low-strain integrity testing will be paid for on a daily-rate basis, as shown in the testing firm's bid documents. The agreed daily rate will be regarded as full compensation for all costs relating to performance of the tests, including preparation and mobilization. Analysis and reporting costs will be paid on an hourly basis, at the rates as shown in the testing firm's bid documents.

(Note: Payment for integrity testing site work on a daily-rate basis is widely regarded in the industry as the fairest method. Payment for analysis and reporting costs can either be on an hourly rate, or on a lump-sum per day of site testing basis. In the event that multiple anomalous results need to be evaluated, the lump-sum basis may not be fair to the testing firm.)

Payment for Shaft-head Preparation

Labor for the preparation of the shaft heads for the integrity testing shall be deemed to be part of the construction cost of the shaft, and will not be paid for as a separate item unless significant backfill and excavation measures are required to expedite the project schedule.

Payment for Core Drilling

In the event that a core is drilled, and the presence of an anomaly is confirmed, the cost of the core drilling will be borne by the foundation contractor. In the event that no anomaly is found at the specified locations, the cost of the core drilling will be borne by the owner and/or the engineer.

2. SAMPLE SPECIFICATION FOR CROSS-HOLE SONIC LOGGING (CSL)

Note: The quantities shown in braces {*nn*} are only recommendations and may be varied to suit the project requirements or engineer's preferences.

OVERVIEW OF CROSS-HOLE SONIC LOGGING (CSL)

The velocity of an ultrasonic pulse through concrete is a function of the modulus and density of the material. The Ultrasonic Pulse Velocity (UPV) is therefore a guide to concrete quality. Material anomalies within the concrete, such as soil inclusions, segregation, voids or honeycombed material, will cause a reduction in pulse velocity and/or amplitude. If multiple measurements are made at different locations, they can provide an assessment of concrete homogeneity or uniformity. The CSL test is a procedure for assessing the structural integrity of a drilled shaft, and identifying the location and extent of any defects detected in the material. This test requires the installation of access tubes for the transmitter and receiver probes. If the tubes are installed in reasonably parallel positions, then the 'time-of-flight' of an ultrasonic pulse between a transmitter probe and a receiver probe in any pair of tubes should be reasonably constant. Any significant variation in pulse travel time or amplitude is potentially indication of an anomaly in the concrete. Careful scrutiny of the CSL test data and review of the construction records by an experienced CSL specialist will be required to determine the likely nature and significance of the anomaly. Subsequent analysis by the geotechnical engineer of record will be required to determine the acceptability of any shaft containing anomalous material.

ACCESS TUBE INSTALLATION

The contractor shall supply schedule 40 mild steel access tubes for the CSL tests. Every shaft drilled under water or slurry ('wet' shafts) shall be equipped with these access tubes. At a minimum, the first {20} % of wet shafts constructed shall be tested by the CSL method. If tests on the first {20} % of shafts show no anomalies that are judged by the engineer to be significant defects, the actual number of the remaining shafts to be tested shall be at the discretion of the engineer, provided that a minimum of {25} % of the total shafts on the project are tested. If anomalies are detected in the first 20 % of shafts that are judged to be significant defects, each subsequent shaft shall be tested until the engineer is satisfied that the cause of the defects has been identified and remedied satisfactorily.

The tubes shall be 38 mm or 50 mm (1.5 or 2 in) internal diameter. (n) access tubes shall be installed in each (n) diameter drilled shaft, equally spaced around the interior of the reinforcing cage. If the tubes are fixed directly to the helical steel or hoops of the reinforcing cage, each tube shall be tied to the cage at intervals not exceeding 3 m (10 ft). If stand-offs or chord supports are used to hold the tubes away from the reinforcing cage, each tube shall be so supported at intervals not exceeding 1.5 m (5 ft).

(Note: The number of tubes is typically based on drilled shaft diameter, with one tube for every 0.3 to 0.35 m (12 to 14 in) of diameter. Most specifications for smaller-diameter shafts stipulate a minimum of three access tubes. If the diameters of the shafts on a given site vary, then the number of tubes must be specified for each shaft size. If several different shaft sizes are required on a project, it may be expedient to include a table that lists shaft sizes and required numbers of tubes.)

Used tubes are acceptable provided that the tubes are clean and regular in cross-section, with no significant rust scale and no internal blockages. The tubes shall be free of oil or grease that would impair the bond between the tube and the concrete. When multiple tube lengths are required, the sections shall be joined with threaded external sleeve couplers to provide a watertight joint with no restriction of the internal diameter. Self-amalgamating tape or another similar sealant may be used on the threads to ensure a watertight seal, but no external wrapping of the joint, such as duct tape, will be permitted. Butt-welding of joints will also not be permitted. Each tube shall have a watertight plug or shoe fitted at its base to prevent ingress of slurry or concrete during construction of the shaft. The tubes shall be installed onto the reinforcing cage in such a manner that the bases of the tubes will be no more than {150 mm (6 in)} above the base of the excavation when the reinforcing cage is installed. The tops of the tubes shall be at least {900 mm (36 in)} above the top of the concrete but no more than {1200 mm (48 in)} above site grade level when the shaft is completed.

(Note: The actual finished level of the tubes will vary according to site-specific requirements, but the practicalities of access and efficient performance of the CSL tests should be considered. A tube 'stick-up' of about 3 ft above the top of the concrete, but no more than 4 ft above grade, is preferred.)

The reinforcing cage shall be handled, picked and placed in such a manner that the integrity and uniformity of the CSL access tubes is not compromised. The CSL access tubes shall be filled with clean non-turbid water immediately after installation of the reinforcing cage or no more than 1 h after completion of concrete placement. The water level in the tubes shall be checked and topped up as necessary at least 30 min, but no more than 2 h, after filling.

(Note: Air bubbles adhering to the walls of the tubes will gradually percolate out after the water is placed. In deep shafts, this can result in a drop in water level of 4 m or more soon after filling.)

Prior to concrete placement, the top of each tube shall be capped in a manner that will prevent concrete or debris falling into the tubes. The capping method shall be one that will not require significant torque, hammering or other mechanical disturbance to uncap or recap the tube after concrete placement.

At the completion of concrete placement, the contractor shall assess the regularity of the access tubes by lowering a 'dummy' representative of a CSL transducer to the bottom of each tube and retrieving the same. In 38 mm (1.5 in) tubes, the dummy shall be of rigid construction, 25 mm (1.0 in) in diameter. In 50 mm (2.0 in) tubes, the dummy shall be 38 mm (1.5 in) in diameter. In either case, the dummy shall be 200 mm (8 in) in length.

If the dummy probe is unable to pass through the full length of any access tube, the contractor shall provide alternative access by core drilling an equivalent-depth access tube in the near vicinity of the blocked or obstructed tube. The core-drilling contractor shall exercise care to avoid cutting or damaging the reinforcing cage. The drilled-shaft contractor shall provide an 'azimuth log' of the core hole so that the CSL testing agency can estimate the location of the core hole relative to the other CSL access tubes at any depth in the shaft. The core drilling shall be monitored and logged by an experienced inspector, to provide a written record of the material removed from the core hole. All costs incurred by any core drilling that is necessitated by obstructed CSL access tubes shall be borne by the drilled-shaft contractor.

Prior to the performance of the CSL tests, the contractor shall provide both the engineer and the CSL consultant with a record of all drilled shaft lengths, plus the elevations of the top and bottom of each shaft, and the date of concrete placement.

(Optional insertion: If CSL testing is completed satisfactorily and the engineer has accepted the shaft, the drilled-shaft contractor shall fill the access tubes completely with cement grout. The grout shall be placed from the bottom up by the use of grout tubes connected to a pump.)

TEST METHOD DESCRIPTION – CROSS-HOLE SONIC LOGGING

The CSL test shall be performed no sooner than {three (3)} calendar days after placement of the concrete in any drilled shaft, and no later than {thirty-six (36)} days after concrete placement on production shafts.

(Note: The testing window may be varied for a test or technique demonstration shaft to suit the project schedule, but if multiple admixtures, such as a plasticizer plus a cement conditioner and/or a retarder are used, the concrete may require more than 3 days to reach a reasonable uniformity – 7 to 10 days is not unusual and longer periods have been reported.)

CSL Test Procedure

The initial set of CSL test profiles shall be developed with the transmitter and receiver transducers in the same horizontal plane in parallel pairs of tubes. Profiles shall be developed for all tube pairs around the perimeter of the shaft, and all diagonally opposed pairs. In shafts with odd numbers of tubes (five or more), profiles shall be developed for the major 'chord' pairs.

Place the CSL transducers in the tube pair to be tested, and lower them to the bottom so that both transducers are at the same elevation. Pull the transducer cables over the measurement wheel, taking CSL measurements every 50 mm or less over the entire concreted length of the shaft. In the event that one tube is shorter than the other, the higher transducer should be held still at the start of the test as the lower transducer is raised until both transducers are at the same elevation, then continue with the test by pulling both transducers simultaneously. The difference in depth must be noted in the report, since it will increase the CSL pulse length at the base of the shaft.

If anomalous data are recorded, and the possibility of equipment malfunction can be eliminated, then additional testing may be required in order to fully evaluate the extent of the anomaly. The engineer should be informed immediately so that the scope of the additional testing can be agreed. If an anomaly is determined to be present in a shaft with more than four access tubes, all possible tube-pair combinations must be tested. In addition, depending on the location of the anomaly, the engineer may require tests with the CSL transducers offset vertically relative to each other so that the signal passes through the anomalous zone at an angle to permit simple tomographic analysis.

The engineer may require additional testing or sampling in order to evaluate the nature and significance of the anomaly. Such additional testing may include Impulse-Echo testing, Impulse-Response Testing, Gamma–Gamma Logging, static or dynamic load testing or core drilling. The actual methods to be used will be determined by the engineer, based on the information required, the access conditions at the shaft and the impact on the project schedule.

Reporting CSL Test Results

A preliminary report shall be presented at the completion of testing, or within one working day of completion of testing. The preliminary report need not include copies of the test data, but must identify which shafts were tested, and whether any shaft contained anomalies which will be discussed in detail in the final report.

A final report shall be presented within {five (5)} working days of the completion of testing. The final report will identify each shaft that was tested, and include the following data:

- Date tested.
- Elevation of top of concrete and/or base of shaft.
- Length of shaft tested.
- Tube-pair orientation diagram with compass reference point.
- CSL data presented as either stacked profile of pulse arrivals versus depth, or computed 'first arrival time' (FAT) and relative pulse energy versus depth.
- Calculated typical pulse velocity across shaft diameter or major 'chord'.
- Calculated % reduction in apparent velocity at any significant anomalies.

The text of the report will include a description of any anomalous zones identified in the test data and a discussion of the apparent difference in pulse velocity. Since the access tubes may move relative to each other during installation and concrete placement, variations in pulse velocity should be assessed on a comparative % basis, using data from a depth that is close to the anomaly. If estimates of actual velocity are to be made, these must be done on diametric profiles, since the diametric pulse length is least affected by tube displacement laterally.

When assessing the likely significance of an anomaly, any available information concerning test cylinders or other actual measurements of concrete strength should be included in the deliberations and recorded in the report. It is not unusual for actual concrete strength to significantly exceed the design strength, in which case the concrete in an anomaly that shows a 15 or 20 % reduction in pulse velocity may still meet or exceed project specifications.

Shaft Acceptance

The engineer shall have {three (3)} working days to evaluate the CSL report and determine whether the tested shafts are acceptable or not. The contractor shall not perform any other work on the tested shafts until the shafts are accepted by the engineer. If the shaft is not acceptable, the engineer will decide which additional testing or investigation methods are necessary in order to accurately characterize the anomaly and determine whether or not the shaft needs to be repaired, and if so, what repair technique will provide satisfactory remediation.

TEST EQUIPMENT REQUIREMENTS

At a minimum, the CSL test equipment shall include:

- A data-acquisition system for recording and display of signals during data acquisition, with an analog-to-digital (A/D) converter capable of at least 12-bit resolution, and a sampling frequency of at least 500 kHz. The data-acquisition system shall record each individual pulse in a format that permits subsequent review and analysis of each pulse, if required.
- Transmitter and receiver transducers which are capable of producing and recording repeatable ultrasonic pulse energy in the frequency range 35 to 50 kHz, with signal amplitude adequate for clear resolution by the data-acquisition system resolution when used in 'good quality' concrete. The transducers shall be sized appropriately to fit in access tubes of 1.5 in (38 mm) minimum internal diameter.
- One or more measurement devices to accurately determine transmitter and receiver probe depths throughout the development of each CSL profile.

PERSONAL QUALIFICATIONS

The CSL consultant shall be an independent testing agency, and the person responsible for analysis and reporting on the CSL test data shall be able to demonstrate a minimum of {five (5)} years' experience in CSL testing, with at least {three (3)} CSL projects in each of those years. The consultant's qualifications and the specifications for the equipment to be used shall be submitted to the engineer for approval prior to beginning drilled shaft installation.

Field personnel responsible for performing or supervising the performance of the CSL test shall be able to demonstrate at least {two (2)} years' experience with CSL testing, with a minimum of {three (3)} projects in each of those years.

BASIS FOR PAYMENT

Payment for CSL Testing

The CSL testing will be paid for on a daily-rate basis, as shown in the testing firm's bid documents. The agreed daily rate will be regarded as full compensation for all costs relating to performance of the tests, including preparation and mobilization. Analysis and reporting costs will be paid on an hourly basis, at the rates as shown in the testing firm's bid documents.

(Note: Payment for CSL site work on a daily-rate basis is widely regarded in the industry as the fairest method. Payment for analysis and reporting costs can either be on an hourly rate, or on a lump-sum per day of site testing basis. In the event that multiple anomalies need to be evaluated, the lump-sum basis may not be fair to the testing firm.)

Payment for CSL Tubes

Labor for the installation of the CSL tubes will be paid for on a lump-sum per shaft basis, as set out in the contractor's bid documents. The cost of material will be paid for on a linear-foot-per-tube basis.

Payment for Core Drilling

In the event that a core is drilled and the presence of an anomaly is confirmed, the cost of the core drilling will be borne by the foundation contractor. In the event that no anomaly is found at the specified locations, the cost of the core drilling will be borne by the owner and/or the engineer.

References

ADSC-IAFD (2005). 'Loadtest, Inc. sets World Record 31 350 Ton Static Load Test', *Foundation Drilling*, **XXVI** (4), May, 57.

ADSC-IAFD/DFI International (2003). *Drilled Shaft Inspector's Manual,* 2nd edn, joint publication by ADSC-IAFD, Dallas, TX, USA and DFI International, Hawthorne, NJ, USA.

American Concrete Institute (1998). *Nondestructive Test Methods for Evaluation of Concrete in Structures*, ACI Report 228.2R-98, ACI, Farmington Hills, MI, USA.

Amir J. (2001). 'Reflections on Pile Integrity Testing', in *Deep Foundations Institute Specialty Seminar on Nondestructive Testing for Drilled Shafts*, St Louis, MO, USA, October 3, DFI, Hawthorne, NJ, USA.

Ang A.H.S. and Tang W.H. (1975). *Probability Concepts in Engineering Planning and Design*, Vol. 1 – *Basic Principles*, John Wiley & Sons, Inc., New York, NY, USA.

Anon (1975). 'Discussion', *Proceedings of the Institution of Civil Engineers*, **59**(2), 867–875.

Arup (2000). [Internet Project Website; www.arup.com/millenniumbridge/index.hml], Arup, London, UK.

ASTM (2000a). 'Standard Test Method for Low-Strain Integrity Testing of Piles', ASTM D5882-00, American Society for Testing of Materials, West Conshohocken, PA, USA.

ASTM (2000b). 'Standard Test Method for High-Strain Dynamic Testing of Piles', ASTM D4945, American Society for Testing of Materials, West Conshohocken, PA, USA.

ASTM (2000). 'Standard Test Method for Integrity Testing of Deep Foundations by Cross-Hole Testing', ASTM D6760, American Society for Testing of Materials, West Conshohocken, PA, USA.

Baker C.N. Jr and Khan F. (1971). 'Caisson Construction Problems and Correction in Chicago', *ASCE Journal of Soil Mechanics and Foundations Division*, **97**(SM2), 417–440.

Baker C.N. Jr, Drumwright E.E., Briaud J-L., Mensah-Dumwah F. and Parikh G. (1993). *Drilled Shafts for Bridge Foundations*, FHwA Publication Number FHwA-RD-92-004, Federal Highway Administration (FHwA), Washington, DC, USA.

Barley A.D. and Woodward M.A. (1992). 'High Loading of Long Slender Minipiles', in *Piling: European Practice and Worldwide Trends,* Institution of Civil Engineers, London, UK.

Beim J., Grävare C-J., Klingmüller O., De-Qing L. and Rausche F. (1998). 'Standardization and Codification of Dynamic Pile Testing – A Worldwide Review', in *Proceedings of the*

Nondestructive Testing of Deep Foundations B.H. Hertlein and A.G. Davis

Deep Foundations Institute Seventh International Conference and Exhibition on Piling and Deep Foundations, Vienna, Austria, 7–9 Oct. 1998, DFI, Hawthorne, NJ, USA.

Benslimane A., Juran I. and Bruce D.A. (1997). 'Group and Network Effect in Micropile Design Practice', in *XIV International Conference on Soil Mechanics and Foundation Engineering*, Vol. 2, P. Seco e Pinto (Ed), Hamburg, Germany, 6–12 Sept, Taylor & Francis, Leiden, NL, 767–770.

Bobrowski J., Bardhan-Roy B.K., Magiera R.H. and Lowe R.H. (1975). 'The Structural Integrity of Large Diameter Bored Piles', in *Conference on Behaviour of Piles: Proceedings of the Institution of Civil Engineers*, ICE, Thomas Telford. London, UK.

Bracewell R.N. (1986). *The Fourier Transform and its Applications,* 2nd edn, McGraw Hill, New York, NY, USA.

Brettman T., Berry W. Jr and Frank M. (1996). 'Evaluation of Defect Detection During Pile Curing Using Sonic Integrity Logging Methods', in *Proceedings of StressWave '96 – Fifth International Conference on the Application of Stress-wave theory to Piles*, F.C. Townsend, Hussein, M., and McVay, M.C. (Eds), Orlando, FL, USA, September 11–13, pp. 688–697.

Briard M. (1970). 'Contrôle des pieux par la méthode des vibrations (Pile Control by the Vibration Method)', *Annales de l'Institut Technique du Bâtiment* (*France*), 23rd year, **270**, June, 105–107 (in French).

British Standards Institution (1986). 'British Standard Code of Practice for Deep Foundations', BS 8004, British Standards Institution, London, UK.

Brown D. (1998). 'Statnamic Lateral Load Response of Two Deep Foundations', in *Statnamic Loading Test 1998: Proceedings of the Second International Statnamic Seminar,* O. Kusakabe, Kuwabara, F. and Matsumoto, T. (Eds), Tokyo, Japan, October 28–30, pp. 415–422, A.A. Balkema, Rotterdam, The Netherlands, pp. 415–422.

Brown D. (2003). 'Self Consolidating Concrete', *Foundation Drilling*, Vol. 23, No. 6, August 24–26.

Bruce D.A. (1994). 'Small Diameter Cast-in-Place Elements for Load Bearing and *in Situ* Earth Reinforcement', in *Ground Control and Improvement*, Xanthakos, P.P., Abramson, L.W., and Bruce, D.A. (Eds), John Wiley & Sons, Inc., New York, NY, USA, pp. 406–492.

Bruce D.A. and Juran, I. (1997). 'Drilled and Grouted Micropiles: State of Practice Review', Report, FHWA RD-97-144, Vols I–IV, NTIS, Springfield, VA, USA.

Bruce D.A. and Nicholson P.J. (1989), 'The Practice and Application of Pin Piling', in *Foundation Engineering: Current Principles and Practice, Proceedings of the 1989 ASCE Foundation Engineering Congress*, Kulhawy F.H. (Ed), 25–29 June, Evanston IL, USA. Pub. ASCE, Reston VA, USA, pp. 272–290.

Bruce D.A., Dimillio A.F. and Juran, I. (1997). 'Micropiles: the state of practice Part I: Characteristics, definitions and classifications', *Ground Improvement Journal,* **1**, pp. 25–35.

Cameron G. and Chapman, T. "Quality Assurance of Bored Pile Foundations". Ground Engineering, Vol. 27, No. 2, pp. 35–40, February 2004, London UK. ISSM 00174-4653.

Cameron G., Wamuziri S.C., Smith I.G. and Chapman T.J. (2002). 'Statistical Sampling Schemes for Integrity Testing of Piled Foundations', in *Proceedings of the United Engineering Foundation International Conference on Probabalistics in Geotechnics: Technical and Economic Risk Estimation,* R. Pottler, H. Klapperich and H. Schweiger (Eds), Graz, Austria, September, 15–19, pp. 129–136. ISBN 3-7739-5977-X.

Chan H.F.C., Heywood C. and Forde M.C. (1987). 'Developments in Transient Shock Pile Testing', in *Proceedings of the International Conference on Foundations and Tunnels,* Vol. 1, Forde, M.C. (Ed), London UK, 24–27 March 1987, pp. 245–261, Engineering Technical Press, Edinburgh, Scotland, UK.

Davis A.G. (1981). 'The Nondestructive Testing of Piles by Mechanical Impedance', in *Proceedings of the Tenth International Conference on Soil Mechanics and Foundation*

Engineering, Session 8, Vol. 2, 15–19 June 1981, Stockholm, Sweden, Balkema A.A. and Rotterdam N.L. (Eds).

Davis A.G. (1985). 'Nondestructive Testing of Piles Founded in Glacial Till – Analysis of over 30 Case Histories', in *Proceedings of the International Conference on Construction in Glacial Tills and Boulder Clays*, Vol. 1, Forde, M.C. (Ed), Edinburgh, Scotland, UK, 12–14, ISBN 0-947644-04-0. March, pp. 193–212.

Davis A.G. (1995). 'Nondestructive Evaluation of Existing Deep Foundations', *ASCE Journal of the Performance of Constructed Facilities*, **9**, 57–74.

Davis A.G. (1997a). 'Nondestructive Evaluation Discussion 1', *Foundation Drilling*, Vol. 36, No. 3, pp. 38–40, March/April, 2–3.

Davis A.G. (1997b). 'Evaluation of Deep Foundations beneath Buildings damaged during the 1994 Northridge Earthquake', in *Innovations in Nondestructive Testing of Concrete*, American Concrete Institute Special Technical Publication SP-168, S. Pessiki and L. Olson (Eds), pp. 319–332, ACI, Farmington Hills, MI, USA.

Davis A.G. (1998). 'Assessing the Reliability of Drilled Shaft Integrity Testing', in *Transportation Research Record 1633*, pp. 108–116, Transportation Research Board (TRB) of the National Research Council, Washington, DC, USA.

Davis A.G. (2003). 'The Nondestructive Impulse Response Test in North America: 1985–2001' *NDT&E International*, **36**, 185–193.

Davis A.G. and Dunn C.S. (1974). 'From Theory to Experience with the Nondestructive Vibration Testing of Piles', Paper 7764, *Proceedings of the Institution of Civil Engineers*, **57**(2), 571–593.

Davis A.G. and Guillermain P. (1979). 'Interpretation géotechnique des courbes de réponse d'une excitation harmonique d'un pieu (Geotechnical Interpretation of the Response of a Pile to Harmonic Excitation)', *La Révue Française de Géotechnique* (*France*), 15–21 (in French).

Davis A.G. and Guillermain P. (1980). 'La vibration des pieux: interpretations géotechniques (Pile vibration: geotechnical interpretation)', *Annales de l'Institut Technique du Batiment et des Travaux Publics (l'ITBTP), France*, No. 380, 71–96 (in French).

Davis A.G. and Hertlein B.H. (1991). 'The Development of Nondestructive Small Strain Methods for Testing Deep Foundations', Paper No. 910243, in *Proceedings of the Transportation Research Board (TRB) Annual Conference*, TRB, Washington, DC, USA, January.

Davis A.G. and Hertlein B.H. (1994). 'A Comparison of the Efficiency of Drilled Shaft Downhole Integrity Tests', in *Proceedings of the International Conference on Design and Construction of Deep Foundations*, Vol. III, Orlando FL, 6–8 Dec. pp. 1272–1286, Federal Highway Administration (FHwA), Washington DC., USA.

Davis A.G. and Kennedy J. (1998). 'Impulse Response Testing to Evaluate the Degree of Alkali–Aggregate Reaction in Concrete Drilled Shaft Foundations under Electricity Transmission Towers', SPIE 3398, Paper 21, in *Proceedings of the Conference on Nondestructive Evaluation of Utilities and Pipelines*, Reuter W.G. (Ed), San Antonio, TX, USA, 1 April, pp. 178–185. The International Society for Optical Engineering (SPIE), Bellingham, WA, USA.

Davis A.G. and Robertson S.A. (1975). 'Economic Pile Testing', *Ground Engineering*, **8**(3), 40–43.

Davis A.G. and Robertson S.A. (1976). 'Vibration Testing of Piles', *Structural Engineer*, **54**(6).

DFI International Drilled Shaft Committee (2003). *4.2 – High Strain Dynamic Testing*', Manual for Testing and Evaluation of Drilled Shafts, Deep Foundations Institute, Hawthorne, NJ, USA.

De Josselin De Jong G. (1956). 'Wat gebuert er in de grond tijdenshet heien (What happens in the soil during pile driving)', *De Ingenieur*, (25), (in Dutch).

DFI International (1995). *Pile Inspector's Guide to Hammers,* 2nd edn, DFI International, Hawthorne, NJ, USA.

DFI International (2003). *Augered Cast-In-Place Pile Manual,* 2nd edn, DFI International, Hawthorne, NJ, USA.

DTU (Document Technique Unifié), (1978). *Travaux de Fondations Profondes pour le Bâtiment (Deep Foundation Works for Buildings)*, French Building Code, Document Technique Unifié (DTU) 13.2, Centre Scientifique et Technique du Bâtiment, Paris, France (in French).

Ellway K. (1987). 'Practical Guidance on the use of Integrity Tests for the Quality Control of Cast *in-situ* Piles', in *Proceedings of the International Conference on Foundations and Tunnels*, Vol. 1, Forde, M.C. (Ed), London UK, 24–26 March 1987, pp. 228–234, Engineering Technical Press, Edinburgh, Scotland, UK.

Felice C.W.; Petek K. and Holtz R.D. (2003). 'What do you do with an Anomaly?', *Foundation Drilling,* **23**(6), pp. 18–22.

Fellenius B. (1995). 'Welcome from the Chairman', in *Proceedings of the First International Statnamic Seminar,* Vancouver, British Columbia, 27–30 September, Pub. Berminghammer Foundation Equipment, Toronto ISBN 0-9680570-0-4.

Finno R.J., Popovics J.S., Hanifah A.A., Kath W.L., Chao H-C. and Hu Y-H. (2001). 'Guided Wave Interpretation of Surface Reflection Techniques for Deep Foundations', *Italian Geotechnical Journal,* Vol. 34(1) 2001, pp. 76–91.

Fleming W.G.K. (1987). 'Quality Assurance in Piling', in *Proceedings of the International Conference on Foundations and Tunnels,* Vol. 1, Forde, M.C. (Ed), 12–14 March 1981 University of London ISBN 0947644-06-7, pp. 128–132, Engineering Technical Press, Edinburgh, Scotland, UK.

Fleming W.G.K., Weltman A.J., Randolph M.F. and Elson, W.E. (1985a). *Piling Engineering*, 1st edn, 1985; 2nd edn, 1992, Surrey University Press, Guildford, Surrey, UK.

Fleming W.G.K., Reiding F. and Middendorp P. (1985b). 'Faults in Cast-in-Place piles and their Detection', in *Proceedings of the Second International Conference on Structural Faults and Repair*, M.J. Forde (Ed.), University of London, UK, pp. 301–310, Institution of Civil Engineers, London, UK.

Forde M.C., Whittington H.W., Coghill G.G. and Batchelor A.J. (1983). 'Electronic Developments in Nondestructive Testing', in *Proceeding of the International Conference on Structural Faults – 83*, Forde M.C., Whittington H.W. and Whyte, I.L. (Eds), Edinburgh, July 1983, pp. 74–85, Engineering Technical Press, Edinburgh, Scotland, UK.

Forde M.C., Chan H.F.C. and Batchelor A.J. (1985). 'Acoustic and Vibration NDT Testing of Piles in Glacial Tills and Boulder Clay', in *Proceedings of the International Conference on Construction in Glacial Tills and Boulder Clays*, Vol. 1, Forde, M.C. (Ed), Edinburgh, Scotland, UK, 12–14 March, pp. 243–256.

Frazier J., Likins G., Rausche F. and Goble G. (2002). 'Improved Pile Economics; High Design Stresses and Remote Pile Testing', in *Proceedings of the Deep Foundations Institute 27th Annual Conference on Deep Foundations*, Weaver, T., Crennan, K. and Pace, A. (Eds), San Diego, CA, USA, 9–11 October, pp. 169–175, DF, Hawthorne, N.J. USA.

Gardner R.P.M. and Moses G.W. (1973). 'Testing Bored Piles formed in Laminated Clays', *Civil Engineering and Public Works Review*, **68**, 60–83.

Gates M. (1957). 'Empirical Formula for Predicting Pile Bearing Capacity', *Civil Engineering*, **27**, 65–66.

Glanville W.H., Grime G., Fox E.N. and Davies W.W. (1938). *An Investigation of the Stresses on Reinforced Concrete Piles During Driving*, Technical Paper No. 20, United Kingdom Department of Scientific and Industrial Research, Building Research, Watford, UK.

Goble G.G., Scanlan R.H. and Tomko J.J. (1967). 'Dynamic studies on the bearing capacity of piles', *Highway Research Record*, **167**, 46–47.

Goble G.G. and Rausche F. (1986). *Wave Equation Analysis of Pile Driving – WEAP86 Program*, Vols I–IV, US Department of Transportation, Federal Highway Administration, Implementation Division, McLean, VA, USA.

Goble G.G., Likins G. and Rausche F. (1975). *Bearing Capacity of Piles from Dynamic Measurements, Final Report*, Research Report, Case Western Reserve University, Cleveland, OH, USA.

Goble G.G., Rausche F. and Likins G. (1980). 'The Analysis of Pile Driving – A State of the Art', in *Proceedings of the (First) International Conference on the Application of Stress-Wave Theory to Piles*, H. Bredenberg (ED.), Stockholm 4–5 June 1980, pp. 131–161, Ashgate Publishing, Stockholm, Sweden.

Guillermain P. (1979). 'Contribution à l'interprètation géotechnique de l'essai d'impédance mécanique d'un pieu (Contribution to the geotechnical interpretation of the mechanical impedance test on a pile)', *Ph.D. Thesis*, Université Pierre et Marie Curie, Paris, France.

Gularte F.B., Taylor, G.E. and Gulorte G.G. (1996). 'Seismic Retrofit of the Fourth Street Viaduct with High Capacity Small Diameter Tiedown Piles', in Soil Improvement for big digs – Proc. Conf. ASCE, 18–21 Oct, Boston MA. ASCE Reston VA., USA. pp. 313–325.

Hannifah A. (1999). 'A Theoretical Evaluation of Guided Waves in Deep Foundations', *Ph.D. Dissertation*, Northwestern University, Evanston, IL, USA.

Hannigan, P.J. (1990). *Dynamic Monitoring and Analysis of Pile Foundation Installation*, Deep Foundations Institute Short Course Text, 1st edn, Deep Foundations Institute, Hawthrone, NJ, USA.

Haramy K.Y. and Mekic-Stall N. (2000). *Crosshole Sonic Logging and Tomographic Imaging Survey to Evaluate the Integrity of Deep Foundations – Case Studies*, Federal Highway Administration, Central Federal Lands Highway Division, Lakewood, Co, USA.

Hayes J. and Simmonds A. (2002). 'Interpreting Strain Measurements from Load Tests in Bored Piles', in *Proceedings of the Ninth International Conference on Piling and Deep Foundations*, 3–5 June 2002, Deep Foundations Institute (DFI International), Hawthorne, NJ, USA, pp. 663–669.

Hayward Baker (1996). *Minipile*, in-house publication, Hayward Baker Corporate Office, Odenton, MA, USA.

Hearne T.M. Jr, Stokoe K.H. II and Reese L.C. (1981). 'Drilled Shaft Integrity by Wave Propagation Method', *Journal of Geotechnical Division*, **107** (GT10), 1327–1344.

Heritier B. and Paquet J. (1986). 'Battage d'un Pieu en Milieu Pulvérulent: Acquisition des Données et Simulation (Pile Driving in Cohesionless Soils: Data Acquisition and Simulation)', *Annales de l' Institut Technique du Bâtiment* (*France*), 39th year, **450**, 1–27 (in French).

Heritier B, Paquet J. and Stain R.T. (1991). 'The Accuracy and Limitations of Full Scale Dynamic Shaft Testing', in Session on Nondestructive Testing of Foundations, presented in Proceedings of the *Transportation Research Board (TRB) Annual Conference*, Washington, DC, USA.

Hertlein B.H. (2004a). Personal correspondence with David Redhead of BSP International Foundations, Ipswich, UK.

Hertlein B.H. (2004b). Personal communication with Brian Liebich of CALTRANS, Sacramento, CA, USA.

Hertlein B.H. (2004c). Personal correspondence with Yajai Tinkey of Olson Engineering, Golden, Co, USA and Zaw-Zaw Aye of SEAFCO, Bangkok, Thailand.

Higgs J.S. and Robertson S.A. (1979). 'Integrity Testing of Concrete Piles by the Shock Method', *Concrete*, **13**(10), 31–33.

Hirsch T.J., Lowery L.L. and Carr L. (1976). *Pile Driving Analysis – Wave Equation User's Manual*, TTI Program Report FHWA-IP-76-14 (4 Vols), US Department of Transportation, Federal Highway Administration, Washington, DC, USA.

Holeyman A., Legrand C., Lousberg E. and D'Haenens A. (1988). 'Comparative Dynamic Pile Testing in Belgium', in *Proceedings of the Third International Conference on the Application of Stress-Wave Theory to Piles*, B.H. Fellenius (Ed.), Ottawa, Can. 25–27 May, pp. 542–554, Bi-Tech Publishers, Ottawa, Canada.

Hong Kong Housing Authority (2000). 'Report of the Investigation Panel on Accountability (Piling Contract 166/1997 Shatin 14B Phase 2)', Hong Kong Housing Authority, [Internet Website: http://www.housingauthority.gov.hk/eng/ha/piling/html/app1_1.htm].

Honma M., Sakai T., Murakumi H., Koyama N., Inagawa H., Miyasaka T. and Tanaka Y. (1991). 'Shape Estimation Method for Cast *in situ* Piles Based on the Stress Wave Theory', in *Proceedings of the Fourth International Conference on Piling and Deep Foundations*, Vol. 1, Stresa, Italy, 7–12 July. Balkema, A.A. Rotterdam NL, pp. 595–600.

Hussein M., Likins G. and Rausche F. (1996). 'Selection of a Hammer for High-Strain Dynamic Testing of Cast-in-place Shafts', in *Proceedings of Stress Wave '96 – Fifth International Conference on the Application of Stress-Wave Theory to Piles*, F.C. Townsend, Hussein, M. and McVay, M.C. (Eds) pp. 759–772, Orlando, FL, USA, 11–13 September, Pub. ASCE, Reston, VA, USA.

Institution of Civil Engineers (1974). 'Esso's Giant Oil Tanks – A Question of More Haste, Less Speed'. *New Civil Engineer*, No. 81 (28 February), 28–38.

Institution of Civil Engineers (1988). *Specification for Piling*, Thomas Telford Ltd, London, UK.

Isaacs D.V. (1931).'Reinforced Concrete Pile Formulae', *Journal of the Institution of Engineers Australia*, **3**, 305–323.

Iskander M., Roy D., Early C. and Kelley S. (2001). 'Class-A Prediction of Construction Defects in Drilled Shafts', *Transportation Research Record 1772*, Paper No. 01-0308, pp. 73–83, Transportation Research Board (TRB), Washington, DC, USA.

Juran I., Benslimane A. and Bruce D.A. (1996). 'Slope Stabilization by Micropile Ground Reinforcement', in *Proceedings of the International Symposium on Landslides*, Senneset, K. (Ed), Trondheim, Norway, 17–21 June Pub. Balkema, Rotterdom NL.

Kilkenny W.M., Lilley D.M. and Akroyd R.F. (1988). 'Steady State Vibration Testing of Piles with Known Defects', in *Proceedings of the Third International Conference on the Application of Stress-Wave Theory to Piles*, B.H. Fellenius (Ed.), Ottawa, Can., 25–27 May, pp. 91–98, Bi-Tech Publishers, Ottawa, Canada.

Kirsch F. and Plassman B. (2002). 'Dynamic Methods in Pile Testing: Developments in Measurement and Analysis', in *Proceedings of the International Deep Foundations Congress*, GeoInstitute, ASCE Geotechnical Special Publication, O'Neill, M.W., Townsend F.C. (Eds), Orlando, FL. 14–16 Feb Reston VA, USA, ASCE, USA.

Kusakabe O., Kuwabara F. and Matsumoto T. (2000). 'Statnamic Loading Test '98', in *Proceedings of the Second International Statnamic Seminar*, Kusakabe, O., Kuwabara, F. and Matsumotu, T. (Eds), Tokyo, Japan, 28 – 30 October, A.A. Balkema, Rotterdam, The Netherlands.

Kwan A.K.H., Zheng W. and Ng, I.Y.T. (2005). 'Effects of Shock Vibration on Concrete', *ACI Materials Journal,* **102**, 405–413.

Leggatt A.J. and Bratchell G.E. (1973). 'Submerged Foundations for 100 000 Ton Oil Tanks', *Proceedings of the Institution of Civil Engineers, Part 1*, **54**, 291–305.

Levy J.F. (1970). 'Sonic Pulse Method of Testing Cast in Situ Piles', *Ground Engineering,* **3**(3), 17–19.

Likins G., Rausche F., Thendean G. and Svinkin, M. (1996). 'CAPWAP Correlation Studies', in *Proceedings of Stresswave'96 – The Fifth International Conference on the Application of Stress-Wave Theory to Piles*, F.C. Townsend, Hussein, M. and McVay, M.C. (Eds), Orlando, FL, USA, 11–13 September, pp. 447–464.

Lilley D.M., Kilkenny W.M. and Akroyd R.F. (1987). 'Investigation of Integrity of Pile Foundations using a Vibration Method', in *Proceedings of the International Conference on*

Foundations and Tunnels, Forde, M.C. (Ed), University of London, 24–26 March pp. 177–183, Engineering Technical Press, Edinburgh, Scotland, UK.

Long J.H., Bozkurt D., Kerrigan J.A. and Wysockey M.H. (1999). 'Value of Methods for Predicting Axial Pile Capacity', Transportation Research Record #1663, *Pile Setup, Pile Load Tests and Sheet Piles*, pp. 57–63, Transportation Research Board, Washington, DC, USA.

Lowery L.L., Hirsch T.J., Edwards, T.C., Coyle H.M. and Samson C.H. (1969). *Pile Driving Analysis – State of the Art*, Research Report 33–13, Texas Transportation Institute, College Station, TX, USA.

McCavitt N. and Forde M.C. (1990). *'Dynamic Stiffness and Effective Mass Parameters of Bored Cast-in-Situ Concrete Piles'*, Research Report, Department of Civil Engineering and Building Science, Edinburgh University, Scotland, UK.

McVay M., Putcha S., Consolazio G. and Alvarez V. (2004). 'Development of a Wireless Monitoring System for Pile Driving', in *Proceedings of the Transportation Research Board (TRB) Annual Conference*, Washington, DC, USA, 11–15 January.

Mitchell J.K. and Villet, W. (1987). (a) *NSHRP Report No. 290*, Transportation Research Board, Washington DC. (1987); (b) *Ground Improvement, Ground Reinforcement, Ground Treatment Development 1987–1997*, ASCE Geotechnical Special Publication No. 69, R. Schaefer (Ed.), ASCE, Reston, VA. (1997).

Middendorp P. and Reiding F.J. (1988). 'Determination of Discontinuities in Piles by TNO Integrity Testing and Signal Matching Techniques', in *Proceedings of the Third International Conference on Application of Stress Wave Theory to Piles*, B.H. Fellenius (Ed.), Ottawa, Can. 25–27 May, pp. 33–43, Bi-Tech Publishers, Ottawa, Canada.

Middendorp P., Bermingham P. and Kuiper B. (1992). 'Statnamic Load Testing of Foundation Pile', in *Proceedings of the Fourth International Conference on Application of Stress-Wave Theory to Piles,* Barends, F.J. and Balkema A.A. (Eds), The Hague, The Netherlands, Rotterdam, pp. 581–588.

Mu R. and Zhao C-S. (1991). 'A New Method for Pile Quality Detection: Estimation of the Shape of a Pile', in *Modal Analysis, Modeling, Diagnostics and Control – Analytical and Experimental*, DE-Vol. 38, Yee, E., and Tsuei, Y.G. (Eds), pp. 135–140, American Society of Mechanical Engineers, New York, USA.

Mullins G. (2004). 'Innovative Load testing Systems, Sub-group Statnamic Testing, Critical Evaluation of Statnamic Test Data' National Cooperative Highway Research Program (NCHRP) Report on Project NCHRP 21–08, Washington DC, USA.

Mullins G., Lewis C. and Justason M. (2002). 'Advancements in Statnamic Data Regression Techniques', in *Deep Foundations 2002: An International Perspective on Theory, Design, Construction, and Performance*, Vol. II, GSP No. 116, Reston, VA. pp. 915–930, ASCE Geo Institute, USA.

NCHRP (2004). 'Load and Resistance Factor Design (LRFD) for Deep Foundations', NCHRP Report No. 507, National Cooperative Highway Research Program, Transportation Research Board of the National Academy of Sciences, Washington, DC, USA.

New Civil Engineer (1974). 'Esso's Giant Oil Tanks – a question of more haste, less speed', *New Civil Engineer*, **81**(February), 28–38.

Niederleithinger E. and Taffe A. (2003). 'Concept for reference pile testing sites for the development and improvement of NDT-CE', in *Proceedings of the International Symposium (NDT-CE 2003):* Non-Destructive Testing in Civil Engineering 2003, 16–19 September, Federal Institute for Materials Research (BAM), Berlin, Germany.

Olson L.D. (2001). *'Determination of Unknown Subsurface Bridge Foundations'*, NCHRP Project 21-5, Preliminary Report, Transportation Research Board National Cooperative Highway Research Program (NCHRP), Washington, DC, USA.

Olson L.D. and Wright C.C. (1989). 'Nondestructive Testing of Deep Foundations with Sonic Methods', in *Foundation Engineering: Current Principles and Practices*, Vol. 2, Kulhawy, F.H. (Ed), pp. 1173–1183, American Society of Civil Engineers (ASCE), Reston, VA, USA.

O'Neill M.W. and Person G.J. (1998). *Innovative Method for Evaluating Drilled Shaft Foundations for the St. Croix River Bridge*, Transportation Research Record 1633: Liquefaction, Differential Settlement and Foundation Engineering, pp. 84–93, Transportation Research Board, Washington, DC, USA.

O'Neill M.W. and Reese L.C. (1999). *Drilled Shafts: Construction Procedures and Design Methods*, Publication No. FHWA-IF-99-025, Chapter 14, pp. 386–422, United States Department of Transportation, Federal Highway Administration, McLean, VA, USA.

Osterberg J.O. (1992). *The Osterberg Load Cell for Testing Drilled Shafts and Driven Piles*, Report to the Federal Highway Administration, Washington, DC, USA.

Paikowsky S.G. and Stenersen K.L. (2000). 'The Performance of the Dynamic Methods, their Controlling Parameters and Deep Foundation Specifications (Keynote Lecture)', in *Proceedings of Stress Wave' 2000 – The Sixth International Conference on the Application of Stress-Wave Theory to Piles*, Niyama, S. and Beim J. (Eds), Sao Paulo, Brazil, 11–13 September, pp. 281–304.

Paikowsky S.G., Regan J.E. and McDonnell J.J. (1994). *A Simplified Field Method for Capacity Evaluation of Driven Piles*, Publication No. FHWA-RD-94-042, US Department of Transportation, Federal Highway Administration, McLean, VA, USA.

Paquet J. (1968). 'Étude Vibratoire des Pieux en Béton: Réponse Harmonique (Vibration Study of Concrete Piles: Harmonic Response)', *Annales de l'Institut Technique du Batîment (Paris, France)*, 21st year, **245**, 789–803 (in French) (English translation by Yee X. (1991), in *Master of English Report*, pp. 33–77, University of Utah, Salt Lake City, UT, USA).

Paquet J. (1969). 'Contrôle des Pieux par Carrotage Sonique (Testing of Piles by Sonic Coring)', *Annales de l'Institut Technique du Batîment (Paris, France)*, 22nd year, (in French).

Paquet J. (1987). 'Évaluation de la Force Portante par Essais Dynamiques (Evaluation of Bearing Capacity by Dynamic Testing)', in *Proceedings of the Belgium Symposium on Pile Dynamic Testing: Integrity and Bearing Capacity*, Brussels, 23–24 Nov. pp. v7–v25, Belgium Society for Soil Mechanics, Brussels, Belgium (in French).

Paquet J. (1988). 'Checking Bearing Capacity by Dynamic Loading: Choice of a Method', in *Proceedings of the Third International Conference on Application of Stress Wave Theory to Piles*, B.H. Fellenius (Ed.), Ottawa, Can. 25–27 May, pp. 383–398, Bi-Tech Publishers, Ottawa, Canada.

Paquet J. (1991), 'A New Method for Testing Integrity of Piles by Dynamic Impulse: The Impedance Log', in *Proceedings of the International Colloquium on Deep Foundations*, École des Ponts et Chaussées, Paris, France, March, pp. 1–10, (in French).

Paquet J. (1992), 'Pile Integrity Testing – the CEBTP Reflectogram', in *Piling Europe: Proceedings of the Conference of the Institution of Civil Engineers*, Sands, M.J. (Ed), London, UK, 7–9 April, pp. 177–188.

Paquet J. and Briard M. (1976). 'Contrôle Non Destructif des Pieux en Béton (Nondestructive Control of Concrete Piles)', *Annales de l'Institut Technique du Bâtiment (Paris, France)*, 29th year, **337**, 50–79 (in French).

Petek K. (2001). 'Capacity Analysis of Drilled Shafts with Defects', *MSCE Thesis,* University of Washington, Seattle, WA, USA.

Petek K., Felice C.W. and Holtz R.D. (2002). 'Capacity Analysis of Drilled Shafts with Defects', in *Proceedings of Deep Foundations 2002, An International Perspective on Theory, Design, Construction and Performance*, Geotechnical Special Publication No. 116, O'Neill, M.W. and Townsend, F.C. (Eds), 14–16 February, American Society of Civil Engineers, Orlando, FL, USA.

Preiss K. (1971). 'Checking of Cast-in-Place Concrete Piles by Nuclear Radiation Methods', *British Journal of Nondestructive Testing*, **13**(3).

Preiss K. and Caiserman A. (1975). 'Nondestructive Integrity Testing of Bored Piles by Gamma Ray Scattering', *Ground Engineering*, **8**(3), 44–47.

Preiss K. and Shapiro J. (1979). Statistical Estimation of the Number of Piles to be Tested on a Project. RILEM Commission on Non-Destructive Testing, Stockholm.

Preiss K., Weber H. and Caiserman A. (1978). 'Integrity Testing of Bored Piles and Diaphragm Walls', *The Civil Engineer in South Africa*, **20**(8), 191–198.

Press F., Ewing W.M. and Jardetzky W.S. (1957). *Elastic Waves in Layered Media*, McGraw-Hill, New York, NY, USA.

Presten M. and Kasali G. (2002). 'Pier Testing Program at University of California Berkeley Site Featuring Conventional Static Load Tests and Rapid Load Tests using the Fundex PLT', in *Deep Foundations Institute: Proceedings of the 27th Annual Conference on Deep Foundations*, San Diego, CA, USA, DFI, Hawthorne, N.J. USA.

Rausche F. and Goble G. (1979). 'Determination of Pile Damage by Top Measurements', in *Proceedings of the ASTM Symposium on Behavior of Deep Foundations*, Boston, MA, June (1978); American Society for Testing and Materials (ASTM) Special Technical Publication, STP 670, pp. 500–506, ASTM, West Conshohocken, PA, USA.

Rausche F. and Robinson B. (2000). 'Newton's Apple Falls in Amherst', *GRL Newsletter*, No. 38 (December), GRL Engineers, Cleveland, OH, USA.

Rausche F. and Seidel J. (1984). 'Design and Performance of Dynamic Tests of Large Diameter Drilled Shafts', in *Proceedings of the Second International Conference on the Application of Stress-Wave Theory to Piles*, C.J. Gravare, G. Holm and H. Bredenberg (Eds), Stockholm, Sweden, 27–30 May, pp. 9–16, Balkema AA, Stockholm, Sweden.

Rausche F. and Seitz J. (1983). 'Integrity Testing of Shafts and Caissons', in Specialty Session on Shafts and Caissons, *ASCE Annual Convention*, Goble, G.G. (Ed), Philadelphia, PA, USA, 16–19 May, PREPRINT 83-033, American Society of Civil Engineers, Reston, VA, USA.

Rausche F., Goble G. and Moses F. (1972). 'Soil Resistance Predictions from Pile Dynamics', *ASCE Journal of Soil Mechanics and Foundations*, 98(SM9), pp. 418–440.

Rausche F., Goble G. and Likins G. (1985). 'Dynamic Determination of Pile Capacity', *ASCE Journal of Geotechnical Engineering*, **111**, 367–383.

Rausche F., Likins G. and Hussein M. (1988). 'Pile Integrity by Low and High Strain Impacts', in *Proceeding of the Third International Conference on Application of Stress-Wave Theory to Piles*, B.H. Fellenius, (Ed.), Ottawa, Can. 25–27 May, pp. 44–55, Bi-Tech Publishers, Ottawa, Canada.

Reiding F.J., Middendorp P. and van Brederode P.J. (1984). 'A Digital Approach to Sonic Pile Testing', in *Proceedings of the Second International Conference on Application of Stress Wave-Theory to Piles*, C.J. Gravare, G. Holm and H. Bredenberg (Eds), Stockholm, Sweden, 27–30 May, pp. 85–93, Balkema AA, Stockholm, Sweden.

Rix G.J., Jacobs L.J. and Reichert C.D. (1993). 'Evaluation of Nondestructive Test Methods for Length, Diameter and Stiffness Measurements on Drilled Shafts', Paper No. 930620, in *Proceedings of the Transportation Research Board Annual Meeting*, Washington, DC, USA.

Robertson S.A. (1976). 'Vibration Testing', *The Consulting Engineer*, (January), 36–37.

Robertson S.A. (1982a). 'Nondestructive Control of Deep Foundations – New Techniques and Recent Experiences', in *Proceedings of the Regional Symposium on Underground Works and Special Foundations*, Singapore, Malaysia, March.

Robertson S.A. (1982b). 'Integrity and Dynamic Testing of Deep Foundations in SE Asia', in *Proceedings of the Seventh Southeast Asian Geotechnical Conference*, Mcfeat-Smith, I., and Lumb P. (Eds), Hong Kong, 22–26 November, pp. 403–421.

Robinson B., Rausche F., Likins G. and Ealy C. (2002). 'Dynamic Load Testing of Drilled Shafts at National Geotechnical Experimentation Sites', in *Proceedings of the International Deep Foundations Congress*, GeoInstitute, ASCE Geotechnical Special Publication, O'Neill M.W., Townsend F.C. (Eds), 14–16 Feb., ASCE, Orlando, FL, USA.

Samman M. M. and O'Neill M.W. (1997a). 'An Exercise in Sonic Testing of Drilled Shafts for Structural Defects', *Foundation Drilling Magazine*, **36**(1), pp. 11–17.

Samman M.M. and O'Neill M.W. (1997b). 'The Reliability of Sonic Testing of Drilled Shafts', *Concrete International*, **19**(1), 49–54.

Samman M.M. and O'Neill M.W. (1997c). 'Fiber-Optic Inspection of Drilled Shafts'. *Foundation Drilling Magazine*, **36**(7), 16–19.

Samman M.M. and O'Neill M.W. (1997d). 'Concretoscopy for Structure Inspection', *Concrete International*, **19**(10), 63–66.

Schaap L. and de Vos J. (1984). 'The Sonic Pile Test Recorder and its Application', in *Proceedings of the Second International Conference on Application of Stress-Wave Theory to Piles*, C.J. Gravare, G. Holm and H. Bredenberg (Eds), Stockholm, Sweden, 27–30 May, pp. 79–84, Balkema AA, Stockholm, Sweden.

Seidel J. and Rausche F. (1984). 'Correlation of Static and Dynamic Pile Tests on Large Diameter Drilled Shafts', in *Proceedings of the Second International Conference on the Application of Stress-Wave Theory to Piles*, C.J. Gravare, G. Holm and (Eds), Stockholm, Sweden, 27–30 May, Balkema AA, Stockholm, Sweden.

Smith E.A.L. (1960). 'Pile Driving Analysis by the Wave Equation', *Journal of the Soil Mechanics and Foundation Division, ASCE*, **86**(August).

Smith R. (2005). 'Roman Pile Driving, Fact from Fiction', *Deep Foundations,* (Spring), 21–25.

Staab D., Edil T. and Alumbaugh D. (2004). 'Nondestructive Evaluation of Cement-mixed Soil', in *Proceedings of Geosupport 2004, ASCE GeoInstitute Conference,* Turner, J.P. and Maybe, P.W. (Eds), Orlando, FL, USA.

Stain R.T. 1982. 'Integrity Testing', *Parts I and II, Civil Engineering* (April), 53–59 and 71–73 (May).

Stain R.T. (1993). 'Discussion of "Stress-Wave Competition/Making Waves"', *Ground Engineering*, **27**(1).

Stain R.T. and Davis A.G. (1983). 'Nondestructive Testing of Bored Concrete Piles – Some Case Histories', in *Proceedings of NDT '83: First International Conference on Nondestructive Testing*, Forde, M.C., Topping, B.H.V. and Whittington H.W. (Eds), Heathrow, London, UK, ISBN 0-947644-01-6 November, pp. 77–87.

Stain R.T. and Davis A.G. (1989). 'An Improved Method for the Prediction of Pile Bearing Capacity from Dynamic Testing', in *Proceedings of the International Conference on Piling and Deep Foundations*, Vol. 1, Burland J., and Mitchell J., (Eds), London, UK, pp. 429–433, Balkema, AA, Rotterdam, The Netherlands.

Stain R.T. and Williams H.T. (1991). 'Interpretation of Sonic Coring Results: a Research Project', in *Proceedings of the Fourth International Conference on Piling and Deep Foundations*, Stresa, Italy, 7–12 April, Balkema AA, Rotterdam, The Netherlands.

Starke W.F. and Janes M.C. (1988). 'Accuracy and Reliability of Low Strain Integrity Testing', in *Proceedings of the Third International Conference. on Application of Stress-Wave Theory to Piles*, B.H. Fellenius (Ed.), Ottawa, Canada, 25–27 May, pp. 19–32, Balkema AA, Rotterdam, The Netherlands.

Steinbach J. (1971). 'Caisson Evaluation by the Stress Wave Propagation Method', *Ph.D. Thesis*, Illinois Institute of Technology, Chicago, IL, USA.

Swann L.H. (1983). 'The Use of Vibration Testing for the Quality Control of Driven Cast *in-situ* Piles', in *Proceedings of the International Conference on Nondestructive Testing*, Forde, M.C., Topping, B.H.W. and Whittington, H.W. (Eds), London, UK, pp. 113–123.

Thorburn S. and Thorburn J.Q. (1977). *Review of the Problems Associated with the Construction of Cast-in-place Concrete Piles*, CIRIA Report, PG2, Construction Industry Research and Information Association (CIRIA), London, UK.

Turner M.J. (1993). 'Discussion of "Stress-Wave Competition/Making Waves"', *Ground Engineering*, **28** (2).

Turner M.J. (1997). *Integrity Testing in Piling Practice*, Ground Engineering: Report 144. p. 336, Construction Industry Research and Information Association (CIRIA), UK.

van Koten H. (1967). *Spanningen in heipalen voor en tijdens het heien (Stress waves in piles before and during driving)*, CUR Rapport No. 42, CUR, Gouda, The Netherlands (in Dutch).

van Koten H. and Wood W.R. (1987). 'Determination of the Shape of Cast-*in-situ* Foundation Piles using the Sonic Echo Technique', in *Proceedings of the International Conference on Foundations and Tunnels*, M.J. Forde (Ed.), London, UK, 24–26 March, pp. 205–210.

van Koten H., Middendorp P. and van Brederode P. (1980). 'Interpretation of Results from Integrity Tests and Dynamic Load Tests', in *Proceedings of the First International Conference on Application of Stress-Wave Theory to Piles*, H. Bredenberg (Ed.), Stockholm, 4–5 June, Ashgate Publishing, Stockholm, Sweden.

van Weele A.F. (1993). 'Discussion of "Stress-Wave Competition/Making Waves"', *Ground Engineering*, **30**(4).

van Weele A.F., Middendorp P. and Reiding F.J. (1987). 'Detection of Pile Defects with Digital Integrity Testing Equipment', in *Proceedings of the International Conference on Foundations and Tunnels*, M.J. Forde (Ed.), London, UK, Engineering Technics Press, pp. 235–244.

Verduin A. (1956). 'Spanningsmetingen tijdens het heien verricht aan een dreital heipalen voor Pier 1 de Waalhaven (Stresswave measurements during the driving of three piles for Jetty 1 of the Harbor (Rotterdam))', TNO Rapport No. 341, TNO, Delft, The Netherlands (in Dutch).

Volkovoy Y. and Stain R.T. (2003). 'Ultrasonic Crosshole Testing of Deep Foundations – 3D Imaging', in *Proceedings of the International Symposium on Nondestructive Testing in Civil Engineering (2003 (NDT-CE, 2003)*, 16–19 November, BAM, Berlin, Germany.

Weihua Z. (2004). 'The CNIS Mission: China's National Standardization Strategies and Cooperation between CNIS and ASTM', *Standardization News*, 32(7), 34–37.

Wellington A.M. (1892). 'Discussion of 'The Iron Wharf at Fort Monroe, VA" (by J.B. Dunklee)', *Transactions, of ASCE Journal*, **27**(Paper No. 543), 29–137.

Weltman A.J. (1977). *Integrity Testing of Piles: A Review*, CIRIA Report PG4, Technical Guide No. 18, pp. 36–37, Construction Industry Research and Information Association (CIRIA), London, UK.

Wennerstrom G. (2004). 'Achieving Excellence in Canadian Construction', in *Proceedings of the American Society for Quality Conference*, ASQ Design and Construction Division, Toronto, Canada.

Wheeler P. (1992). 'Stress-wave Competition/Making Waves (Results of Delft University Pile Integrity Testing Competition)', *Ground Engineering*, **25**, pp. 25–26, (see also subsequent discussions by Stain (1993), Turner (1993) and van Weele (1993)).

Whitaker T. (1963). 'The Constant Rate of Penetration Test for the Determination of the Ultimate Bearing Capacity of a Pile', *Proceedings of the Institution of Civil Engineers*, **26**, 119–123.

Williams H.T. and Stain R.T. (1987). 'Pile Integrity Testing – Horses for Courses', in *Proceedings of the 1987 International Conference on Piling and Deep Foundations*, Vol. 1, Forde, M.C. (Ed), University of London 24–26 March, pp. 184–191, Engineering Technical Press, Edinburgh, Scotland, UK.

Yee X. (1991). English Translation of Paquet J. (1968), 'Étude Vibratoire des Pieux en Béton: Réponse Harmonique (Vibration Study of Concrete Piles: Harmonic Response)', *Master of English Report*, pp. 33–77, University of Utah, Salt Lake City, UT, USA.

Index